The Revolution of Magellan

Romeu Gaspar

The Revolution of Magellan

How 500-Year-Old Rocket Science Changed the World

Romeu Gaspar [ID]
X&Y Partners
Lisbon, Portugal

ISBN 978-3-032-10796-1   ISBN 978-3-032-10797-8   (eBook)
https://doi.org/10.1007/978-3-032-10797-8

This Springer imprint is published by the registered company Springer Nature Switzerland AG
The registered company address is: Gewerbestrasse 11, 6330 Cham, Switzerland

If disposing of this product, please recycle the paper.

To my family, and to the village

Preface

On the slopes of a volcano and facing the vastness of the Atlantic Ocean, stands the small and unassuming Whaler's Museum. Strewn across three old boat houses, it tells the story of Azorean whale hunting, which endured until the late eighties using the same harpoons and lances popularised in Moby Dick.

Those with weak stomachs should watch out for the uncensored images and videos of whale flensing, which show the large cetaceans lying on ports and beaches while being sliced to extract their valuable blubber. The next room showcases one of the tiny wooden boats used to hunt them, next to accounts from the hunters. They were farmers and herders, who jumped onto these fragile vessels after spotting the spouts of sperm whales in the distance, risking life and limb for some extra pay to make up for the meagre earnings of the land. In 1987, when whale hunting was outlawed in Portugal, most of them emigrated.

The Whaler's Museum makes no attempt to sugarcoat either the suffering of the hunted or the hardships of the hunters. It tells the facts and invites the viewer to form an opinion.

Likewise, history books should be informative, rather than prescriptive. They do not need to be stripped of nuance nor shoehorned into a narrative to make them easy to consume. Good stories stand on their own two feet, not on the author's shoulders.

I am a physicist with a keen interest in the history of science and a deep admiration for Carl Sagan and Simon Singh's talent for turning dense topics into engaging—but accurate—stories. Converting that inspiration into a book was one of the hardest things I have ever done.

The first order of business was to find a story worth telling. My father, Joaquim, a retired naval officer turned historian, was a promising source: many family gatherings had ended with the two of us entrenched in a corner, discussing his research on the mathematical properties and distortions of old nautical charts. Stooping under the weight of borrowed volumes, I set to work. After a few months, I had an outline for a book about cartography.

Suspecting an exceedingly narrow focus for a popular science book, my father put me in touch with Henrique Leitão, a fellow scholar with one foot in academia and one in the real world. "Your choice of topic is certainly dense", he said, "but I don't see how you can make it engaging for a broad audience". After slamming that door, Henrique offered a window: "If I was in your shoes, I would write a book called *The Science of Magellan*, talking about all the scientific achievements underpinning the expedition". The 500th anniversary of Magellan's voyage was still several years away, so I had plenty of time.

Or so I thought. Used to working with others on near-term goals, writing a book alone towards a distant deadline offered a vast open ocean to explore. Initially I stayed close to what I had learned about cartography, explaining why Magellan believed the Moluccas were on the Spanish hemisphere (*How Did Magellan Convince the Spanish King?*), and the apparent contradictions of old nautical charts (*Why Did Nautical Charts Ignore the Earth Was Round?*). Then I ventured into naval science (*How Advanced Were Magellan's Ships?* and *What Navigation Techniques Were Available to Magellan?*). Eventually, my curiosity took me well beyond the years of Magellan's expedition (*How Did 16th Century Exploration Change Food Habits Around the Globe?* and *What Happened to the Natives Encountered by the Fleet?*).

Over time, *The Science of Magellan* morphed into *The Revolution of Magellan*. Using the account of the eventful expedition as a central thread, the book not only goes back in time to explore the science behind the first revolution around the planet, but also forward to reveal the ensuing rupture.

This book leverages nearly 500 academic publications. A sizeable portion of them are only available in Spanish or Portuguese, and I hope this endeavour will bring them closer to an English-speaking audience. I trust readers will also find interest in findings that never quite found their way out of academic journals. Indeed, nearly everywhere my curiosity took me, somebody else had already been there. The only question for which I did not find an answer—how was San Martín, the expedition's astronomer, able to accurately measure longitude?—motivated my contribution to the field, where I took NASA's ephemerides (normally used to plan space missions) and simulated San Martín's measurements (*Were San Martín's Longitude Measurements Accurate?*).

Cátia, my wife, was present at every single step of writing this book, from inception to publishing: no detail was too small to discuss, and no draft was too long to review. Lucas, our son, was born about a year after I started this endeavour. He spent countless hours by my side while I wrote, first playing, then drawing, and finally reading.

From both my parents, Maria João and Joaquim, I inherited the joy of learning and writing. My father was also a reference in several of the book's chapters, a relentless reviewer, and my gateway into the inner workings of 16th-century history. Once there, Henrique Leitão straightened out my priorities and provided an initial batch of resources on Magellan's expedition. José Malhão Pereira answered my bottomless inquiries on navigation, Filipe Castro on ships, and Samuel Gessner on astronomy. Šima Krtalić provided feedback on both style and substance, picking out excessive colloquialisms and proposing *The Short Answer* summaries at the beginning of denser chapters. José Maria Moreno Madrid, José Manuel Garcia, and Teresa Nobre de Carvalho peer reviewed the book, challenging conclusions, proposing better sources, and weeding out anachronisms. At Springer Nature, my editors Angela Lahee and Sam Harrison readily embraced this project and provided invaluable feedback, while Divya Sureshkumar, Vidya Velmurugan, and the rest of the production team took it to the finish line. A final nod to all the institutions making content freely and readily available online, from old documents to the latest academical research: Spain's PARES, Portugal's Biblioteca Nacional Digital, Google Books, Project Gutenberg, the Internet Archive, the Wikipedia Library, and many others.

It took ten years and a village to write this book. I am deeply grateful to all.

A final acknowledgement to you, the reader: with so many stories worth reading about, thank you for choosing this one.

Lisbon, Portugal   Romeu Gaspar

Competing Interests The author has no competing interests to declare that are relevant to the content of this manuscript.

Contents

1

A Smooth Departure

From Sanlúcar de Barrameda (September 20th, 1519) to Tenerife (September 26th, 1519)

Just before dawn, lured out of the harbour by favourable winds rather than the first hints of daylight, a fleet of five ships left Sanlúcar de Barrameda, in southern Spain (Fig. 1.1). Those who watched from the shore, if not privy to the expedition, would not think much of it, as the ships did not stand out in the busy harbour.

A closer inspection, however, would reveal this was no ordinary voyage. The fleet's flagship, although flying the Spanish royal standard, was commanded by Ferdinand Magellan, a Portuguese navigator. Magellan, born into a family of minor nobility and modest possessions, had somehow convinced King Charles V[1] it was possible to reach the eastern riches by travelling west. The goal was to find an alternative path to the Moluccas—the source for cloves, today part of Indonesia—and pull them from under Portugal's control.

It is hard to find a parallel to Magellan's proposal in our times. Those five ships were expected to—in a single journey—find a yet unknown passage from the Atlantic to the Pacific, sail across the vast expanse of the latter, find the miniscule Moluccas, establish local alliances, prove they were in Spain's hemisphere, and then retrace the entire trip back to Spain with a full haul of

---

[1] By the time the fleet departed, the young Habsburg monarch was both Charles I, the king of Spain, and Charles V, the Holy Roman Emperor. Both titles are used throughout the book (the former for events before the coronation, and the latter afterwards).

R. Gaspar, *The Revolution of Magellan*, https://doi.org/10.1007/978-3-032-10797-8_1

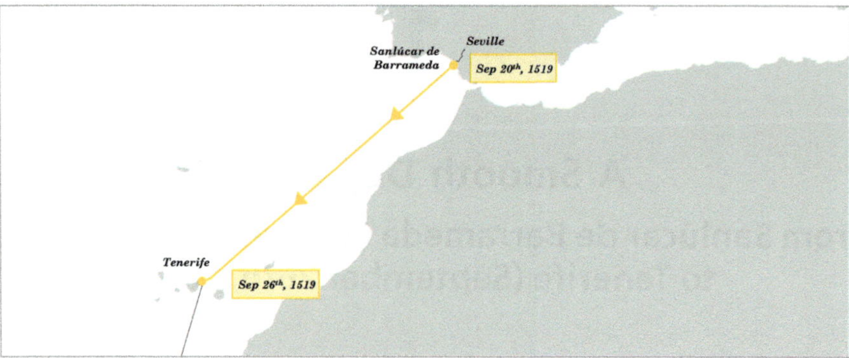

Fig. 1.1 Fleet route[2]

cloves.[3] The likelihood of failing at any one of these steps was immense, but so were the rewards if they succeeded.

In any case, they were off to a good start. Nothing out of the ordinary happened on the first leg of the trip and the fleet arrived at Tenerife after the customary six days, following the same southwestern route well known to Spanish vessels.[4]

1.1 How Did Magellan Convince the Spanish King?

The short answer

In 1494, to avoid overseas conflicts, Spain and Portugal signed the Treaty of Tordesillas, dividing the globe into two hemispheres of influence. Initially the treaty worked well, with the Spaniards duly expanding westward to the Americas and the Portuguese eastward to India. However, in 1512 Portugal reached the Moluccas, sparking complaints from Spain that the prized Moluccas

---

[2] Fleet route adapted from Tomás Mazón Serrano ("Mapas", Ruta Elcano, June 19, 2024, https://rutaelcano.com/mapas/); map adapted from Natural Earth (1:10 m large scale land data, version 5.1.1, https://www.naturalearthdata.com/downloads/).

[3] "Capitulación con Fernando de Magallanes y Ruy Falero encomendándoles la Armada para el descubrimiento de la Especiería" (Seville: Archivo General de Indias, ES.41091.AGI/24//INDIFERENTE,415,L.1,F.18 V-20R, 1518), https://pares. Mcu.es/ParesBusquedas20/catalogo/description/244379; "Instrucción de Carlos I a Fernando de Magallanes y a Ruy Falero, y requerimiento de Fernando de Magallanes a la Casa de la Contratación" (Seville: Archivo General de Indias, ES.41091.AGI//PATRONATO,34,R.8, 1519), https://pares.mcu.es/ParesBusquedas20/catalogo/description/122217.

[4] Antonio Pigafetta, *The first voyage around the world 1519–1522*, edited by Theodore Cachey Jr. (Toronto: University of Toronto Press, 2007), on p. 6.

were too far east and already part of its hemisphere. Due to the difficulties of accurately measuring longitude, the dispute simmered. In 1518, Magellan showed the Spanish king a chart plotting the Moluccas 2° inside the monarch's hemisphere. As the chart was based on Portugal's own cartography, Charles V felt no compunction in charging Magellan with the mission of finding a western route to the Moluccas. Alas, unbeknownst to both men, the small group of islands was still part of Portugal's hemisphere. Portuguese nautical charts, based on compass readings uncorrected for magnetic declination (the difference between a compass reading and true north, caused by fluctuations in the Earth's magnetic field), exaggerated the distance between Europe and the Moluccas, pushing the latter into the Spanish hemisphere. Magellan's entire endeavour had been inspired by an arbitrary arrangement of the Earth's molten metallic core.

By the sixteenth century, clove had already been known in Europe for centuries—its first known written reference dates back to 70 AD[5]—and was firmly embedded in medieval refined cuisine. Like other spices, cloves were largely used to season otherwise dull dishes (lacking ingredients which today are commonplace, such as peppers, tomato, cocoa, vanilla, sugar, and coffee[6]) and to help conserve cooked meat.[7]

Clove initially found its way to Europe through the Silk Road. During the long journey, it would pass through ten hands or more: the slave who plucked the blossoms in the Moluccas; the master that sold them at the local market; the merchant who bought and hauled the produce to Malacca; the Malaccan sultan that demanded tribute for all goods passing through his harbour; the flurry of freighters that, port by port, transported it all the way to the Persian or Egyptian shores; the local emirs and sultans who also demanded a share of the profits; the owner of the camels that took it across the dunes; the Venetians who sailed it from Alexandria to Venice; the German, Flemish, and other traders that bought them from the Venetians and distributed it across Europe; and finally the retailer who sold it to the final wealthy customer. And wealthy were indeed those customers. Much like pepper and other spices, the scarcity and high price of clove made it into a currency and a status symbol, to the point where dishes of the richest could turn almost inedible, so pungent was the seasoning.

---

[5] In Pliny the Elder, *The Natural History of Pliny*, translated by John Bostock and Henry Riley (London: H.G. Bohn, 1855–57) in Book XII, Chap. XV.

[6] See How Did 16th Century Exploration Change Food Habits Around the Globe?

[7] Luís Thomaz, *O drama de Magalhães e a volta ao mundo sem querer* (Lisboa: Gradiva, 2018), on p. 68.

Sensing there was money to be made, in the early fifteenth century the Portuguese started seeking a maritime alternative to the Silk Road. By 1488 they had rounded Africa's southernmost cape, disproving the notion that the Indian Ocean was landlocked, as widely believed across Europe since the ancient Greeks. By 1498 they had reached India; by 1511 they had taken Malacca, the region's most important trading port; and by 1512 they had reached the Moluccas and confirmed they were indeed the source of cloves.

There was now a far more direct way to bring cloves to Europe, by sea rather than by land, and controlled by a single pair of hands, instead of ten. The allure of controlling its supply chain was immense: in years of scarcity, clove could be 240 times more expensive in Lisbon than in the East.[8] Of course, it was naïve for Portugal—a small nation of little more than one million[9]—to assume it could singlehandedly maintain control over such an important commercial route. Indeed, the first attempts to circumvent the new maritime trading route started even before Portugal arrived to India.

In 1493 Christopher Columbus claimed to have reached the Indies— the name given by Europeans to South and Southeast Asia—after sailing west across the Atlantic for a mere five weeks. On the return leg, a storm forced him to stop at Lisbon before continuing to Spain, where King John II summoned him for a hearing. The king was far from thrilled with the expedition, telling Columbus he had breached the Treaty of Alcáçovas. Signed in 1479 between Spain and Portugal, this agreement had ended the War of the Spanish Succession by awarding Portugal with sea rights south of the Canary Islands in exchange for waiving their claims to the Spanish throne.[10]

King John II was not convinced Columbus had reached the Indies.[11] Nevertheless, the expedition strained the diplomatic relations between the two neighbouring countries. The Portuguese king quickly sent a letter to Ferdinand II and Isabella I, the Spanish Catholic monarchs, complaining they had broken the Treaty of Alcáçovas. As a response, the Spanish monarchs brought Pope Alexander VI to the dispute, who quickly weighted in. The pope—Spanish by birth—issued a bull establishing a demarcation line 100

---

[8] Luís Thomaz, "Maluco e Malaca", *Actas do II Colóquio Luso-Espanhol de História Ultramarina*, 1975, p.27–48, on p. 44.

[9] José Machado, "No Centenário do I Recenseamento Populacional Português", *Revista do Centro de Estudos Demográficos*, 16, 1965, p. 83–104, on p.87.

[10] Samuel Eliot Morison, *Admiral of the sea—A life of Christopher Columbus* (Boston: Northeastern University Press, 1983), on p. 197–344.

[11] Idem, on p. 381.

leagues (~590 km[12]) west of the Portuguese islands (the Azores and Cape Verde), and awarding any non-christened lands left of the line to Spain. Confirming the growing suspicions Columbus had not reached the Indies, a second bull followed, explicitly awarding the Indies to Spain irrespective of their position relative to the demarcation line.[13]

The Catholic Church counted Spain and Portugal amidst its closest allies, as both countries had dutifully driven Muslims out of the Iberian Peninsula and vowed to expand Christianity to newfound lands. Still, the Portuguese king's Catholic devotion had its limits. Faced with two papal bulls which ignored the earlier Treaty of Alcáçovas and threatened his maritime route to India, John II sought to negotiate a more advantageous agreement directly with the Spanish monarchs. Knowing the Pope's decision to be heavy-handed and wary of the superior Portuguese naval force,[14] Ferdinand II and Isabela I agreed to sit down with John II to negotiate a new agreement.

The resulting 1494 Treaty of Tordesillas helped to diffuse tensions by shifting the pope's demarcation line a further 270 leagues (~1,600 km) west, giving the Portuguese ships returning from the east ample space to take advantage of the mid-Atlantic north-easterly trade winds (Fig. 1.2). The treaty also omitted any mention of the Indies. This tacitly meant that if Columbus had indeed reached the Indies after crossing the Atlantic, they belonged to Spain. If not, they belonged to Portugal, but Spain could keep Columbus' newfound lands.[15] Like the Treaty of Alcáçovas that preceded it, the agreement penned in Tordesillas was one of several instances where Spain and Portugal replaced war by diplomacy and compromise. The matter was not quite settled, though.

The treaty's demarcation line was a meridian extending from pole to pole, splitting the Atlantic but not reaching the other side of the globe. At the time, sailing there from Europe was still a distant dream, so the omission was not particularly concerning to either signatory. However, that changed over the next decade, as Portugal quickly expanded across Asia. In 1509,

---

[12] Similarly to other old measuring units, the length of a Castilian league cannot be precisely determined. Most estimates place it between 5,573 m (Salvador García Franco, *La Légua Nautica en la Edad Media*, Madrid, Instituto Historico de Marina, 1957, on p. 29) and 5,920 m (Abel Fontoura da Costa, *A Marinharia dos Descobrimentos*, Lisbon, Ministério do Ultramar, 1939, on p.216, Table XX).

[13] Emma Blair and James Robertson, *The Philippine Islands* (Cleveland: A.H. Clark Company, 1903), in Volume 1, p. 97–114.

[14] Bailey Diffie and George Winius, *Foundations of the Portuguese empire 1415–1580* (Minneapolis; University of Minnesota Press, 1985), in Volume I, p. 152.

[15] Emma Blair and James Robertson, *The Philippine Islands* [...], on p. 115–129.

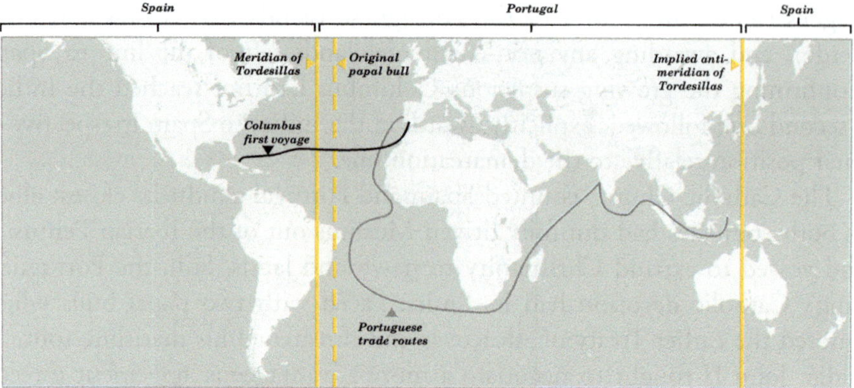

Fig. 1.2 Columbus' first voyage prompted Portugal and Spain to sign the 1494 Treaty of Tordesillas, which divided the globe in two hemispheres of influence and diffused conflict between Portugal's eastward expansion and Spain's western advances

when the country's ships reached Malacca, Spain worried that this important trading port was already east of the anti-meridian demarcation line implicitly assumed in the treaty. The issue became unavoidable in 1512, when Portugal claimed the rich Moluccas.

In which side of the anti-meridian demarcation did the Moluccas fall? Without the means to accurately measure longitude,[16] there was no clear answer. Magellan argued they were part of the Spanish hemisphere, and he was not the only Portuguese thinking so. Francisco Serrão, his close friend, had been one of the first Portuguese to arrive at the Moluccas and had settled there. Serrão had exaggerated the distance he travelled from Malacca to the Moluccas, adding to the suspicions the islands may already be east of the demarcation line.[17]

In his hearing with the Spanish king, Magellan showed letters Serrão had sent him,[18] reporting on the location of the Moluccas and its resources. These letters—plus Magellan's own time spent in the Indies[19]—caused quite an impact, as they provided first-hand experience and intelligence about the islands which Spain lacked. However, if these were Magellan's only pieces of evidence he would probably still have struggled to convince the monarch and his advisors, as his predecessors had.

---

[16] See Were San Martín's Longitude Measurements Accurate?

[17] João de Barros, *Década terceira da Ásia de João de Barros* (Lisbon: Jorge Rodriguez, 1628), in Livro V, f.134.

[18] Idem, on f.141.

[19] See Why Was Magellan, a Portuguese Citizen, Serving Spain?

Columbus had spent the better part of a decade chasing financial support for his endeavour of reaching the Indies by sailing west. His difficulty in finding a patron was not, as is sometimes stated, because people thought the Earth was flat.[20] The planet's sphericity was known at least since Aristotle (384–322 BC), who inferred it from the Earth's circular shadow cast on the Moon during an eclipse and the fact that the constellations visible in the sky changed with latitude.[21] No, Columbus' problem was his unabashed use of doubtful data and questionable logic.

In the late fifteenth century, the European view of the planet was still deeply shaped by the works of Claudius Ptolemy (c.100—c.170 AD), a Greek polymath who had written *Geography*, a review of the Greco-Roman cartographic and geographical body of knowledge. Ptolemy was not trusted blindly (for instance, his belief that the Indian Ocean was landlocked had been disproved in 1487, when Bartolomeu Dias rounded the Cape of Good Hope), but he was the de facto source when more recent intelligence was not available. Namely, Ptolemy's estimate that there was 180° separating the western tip of Europe from *Magnus Sinus* (an amalgamation of the Gulf of Thailand and the South China Sea) was accepted by many, although it overshot the true distance by about 60°.

Still, that was not enough for Columbus, who instead used a figure of 225° sourced from Marinus of Tyre (70–130 AD), a precursor of Ptolemy. He then added 28° for Marco Polo's discoveries, 30° for the China to Japan distance, and an extra 8° to correct for—in his view—Marinus of Tyre's ill assumptions when converting between linear distances and degrees of longitude. This tallied 291°, meaning that Japan was just 69° west of continental Europe (360° minus 291°) or 60° west of the Canary Islands, from where he intended to start his journey. When converting this longitudinal difference into an absolute distance, Columbus made an additional 25% underestimate by assuming Al-Farghani (c.800—c.870 AD) had used smaller Roman miles instead of Arabic miles for his estimate of the perimeter of the Earth.[22]

---

[20] It is hard to precisely pinpoint the origin of this misconception, but Washington Irving's 1828 romanticised biography of Columbus and the 1937 popular song "They All Laughed" surely contributed to it.

[21] Interestingly, Aristotle did not use what is perhaps the most popular empirical argument for a round Earth: ships at a distance dip below the horizon; Dirk L. Couprie, *When the World Was Flat: Studies in Ancient Greek and Chinenese Cosmology* (Cham: Springer, 2018).

[22] Samuel Eliot Morison, *The European Discovery of America - The Southern Voyages* (New York: Oxford University Press, 1974), on p. 65.

Combined, Columbus' bloated Eurasia and smaller Earth placed Japan just some 4,400 km[23] away from the Canaries, less than four times the true distance of 19,600 km. The gross underestimate put the Indies within reach of fifteenth century ships but made his case hard to sell to anyone who investigated the underlying assumptions. After several attempts, Columbus did secure the backing of the Spanish monarchs, who remained sceptical but were afraid the Italian navigator would pitch his plan elsewhere.[24]

After Columbus returned from his first voyage, the Spanish Crown—perhaps more out of self-interest than genuine belief—took the official position that the navigator had indeed reached the Indies.[25] However, the subsequent expeditions made the proposition increasingly hard to defend, feeding suspicions that Columbus had not reached Asia but an unknown vast continent. At this point, Spain's expansion goals forked, and the kingdom set on exploring the Americas while keeping a discreet eye on opportunities to sail further to the west. In 1513, Vasco Núñez de Balboa reached the Isthmus of Panama and spotted the Pacific Ocean (which he called *Mar del Sur*, or Southern Sea), raising the hope that somewhere in the Americas lied a passage to the Indies. In 1514, the Spanish Crown secretly ordered João Dias de Solis, a Portuguese national who had raised to the rank of Pilot Major,[26] to search for a passage south of the Portuguese occupied territories in Brazil. The attempt ended terribly at the mouth of the La Plata River (present-day border of Argentina and Uruguay), when Solis and a small landing party were attacked and, according to survivors, eaten by natives.[27]

So, by Magellan's time, several before him had tried and failed to reach the Indies travelling west, each one putting additional strain on the Treaty of Tordesillas signed with Portugal. The risks and difficulties of such an endeavour had been made clear, and the Spanish Crown was not inclined to finance any more ill-conceived endeavours. Magellan was different though. Not only had he spent time in Asia and had a friend in the Moluccas, but the scientific reasoning of his plan was also more robust and he had partnered with Rui Faleiro, a Portuguese mathematician and astronomer. Faleiro was

---

[23] Assuming 1480 m per Roman mile (Abel Fontoura da Costa, *A Marinharia dos Descobrimentos* [...]).

[24] William Phillips and Carla Rahn, *The Worlds of Christopher Columbus* (Cambridge: Cambridge University Press, 1992).

[25] Samuel Eliot Morison, *Admiral of the sea* [...], on p. 381.

[26] The highest pilot rank in Spain, tasked with the preparation and execution of expeditions.

[27] Antonio Herrera, *Historia general de los hechos de los castellanos en las islas i Tierra Firme del Mar Oceano* (Madrid: En la emplenta Real, 1601), in Decada II, Libro I, Cap. VII, p. 13.

irascible and seemingly mentally unstable—to the point of being excluded from the expedition in its later stages of preparation—but even those who detested him admired his abilities.[28] Magellan certainly regarded him highly, making him an equal partner in the endeavour. Besides vetting the scientific foundation of Magellan's claims, Faleiro provided three methods to determine the longitude of the Moluccas once they reached them.[29] This was a critical point: what good would it be to circle more than half of the world only to be unable to prove the islands to be part of the Spanish hemisphere?

Rather than cherry-picking and distorting estimates from ancient Greeks and Arabs, like Columbus had done, Magellan and Faleiro leveraged up-to-date nautical charts made by the Portuguese, which incorporated the latest findings from their voyages to the Indies. Magellan prepared a memorandum pinpointing the position of the Moluccas and other key locations,[30] and commissioned Pedro and Jorge Reinel (father and son, two renowned 16th Portuguese cartographers) to make a companion chart.[31] This chart, today known as the Kunstmann IV,[32] places the Moluccas about 2° to the east of the Tordesillas anti-meridian, just inside the Spanish hemisphere. Magellan did not sugarcoat the journey though, as the chart showed that a Spanish western pathway to the Moluccas would be longer than the established eastern Portuguese route, against Columbus' claim that it was quicker to reach Asia from the west than from the east.

It is indeed much quicker to reach Asia from the east than from the west. However, and regrettably for Magellan and Charles V, the Kunstmann IV chart got the most important thing wrong: the Moluccas were not 2° east of the Tordesillas anti-meridian but 5° west of it, in the Portuguese hemisphere (Fig. 1.3). The chart shifted the position of the coveted islands by 7°, but why?

The most obvious explanation is that Magellan had doctored the chart to deceive the Spanish king. However, contemporary charts made by the Reinel and other Portuguese cartographers also shifted the Moluccas eastward.[33] The question then becomes even more interesting: why would Portugal place these

---

28 Idem, in Decada II, Libro II, Cap. XIX, p. 64–67.

29 See Were San Martín's Longitude Measurements Accurate?

30 Ferdinand Magellan, *Memorial atribuido a Magallanes en el que se justifica la pertenencia de las Molucas a España* (Seville: Archivo General de Indias, ES.41091.AGI//PATRONATO,34,R.13, 1519), on p. 29, https://pares.mcu.es/ParesBusquedas20/catalogo/description/122222.

31 Dejanirah Couto, "*Les cartographes et les cartes de l'expédition de Fernand de Magellan*", *Anais de História de Além-Mar XX*, 20, October 2019, p. 81–120, on p. 84.

32 Attributed to Jorge Reinel and Pedro Reinel, *Kunstmann IV Nautical Planisphere* (facsimile by Otto Progel) (Paris: Bibliothèque Nationale de France, CPL GE AA-564, c1519).

33 Joaquim Gaspar and Šima Krtalić, *The Cartography of Magellan* (Lisbon: Tradisom, 2023), on p. 138–142.

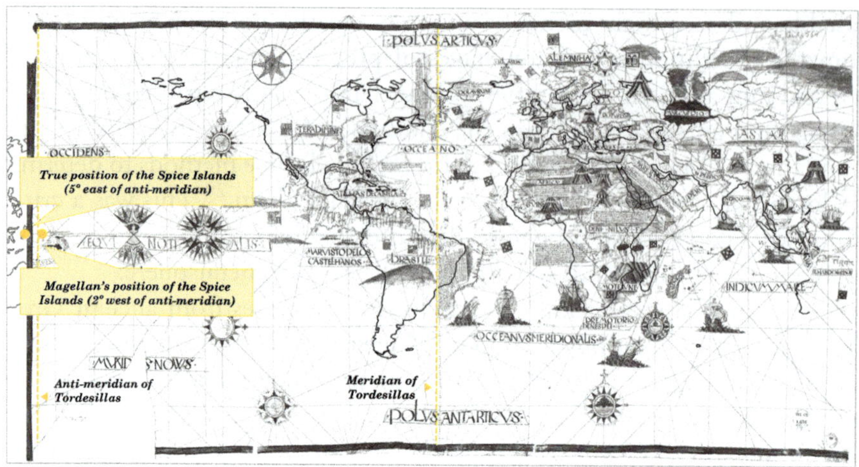

Fig. 1.3 The Kunstmann IV chart overlaid with an accurate representation of the continents, showing the Moluccas to be 5° east of the Tordesillas anti-meridian, not 2° west of it[34]

islands in the Spanish hemisphere, against its own interests? The issue had begun centuries ago, long before Portugal and Spain's maritime expansion, and had remained in the shadows, undetected and inconsequential.

Today, there is no fundamental difference between a map (used to show the location and shape of landmasses) and a chart (used to navigate a ship or a plane), as both are a collection of locations marked over a grid of coordinates. That was not the case in medieval Europe. While maps already used a coordinate system of latitude and longitude, nautical charts were based instead on headings and distances.

These first nautical charts were called portolan charts, and the earliest extant example is the thirteenth century *Carta Pisana*.[35] Portolan charts forwent latitude and longitude because Mediterranean sailors were not able to measure them. Instead, they planned voyages and tracked progress based on heading (the ship's direction, measured with a compass) and distance (estimated from the ship's speed). For instance, a pilot wishing to go from Alexandria to Patara (an ancient maritime city in modern-day Turkey) could consult a portolan chart to know he needed to sail north for about 600 km. Nautical charts were not only a better fit for navigation, but they were

---

[34] Attributed to Jorge Reinel and Pedro Reinel, *Kunstmann IV* […], accurate representation from continents adapted from Natural Earth (1:10 m large scale land data, version 5.1.1, https://www.nat uralearthdata.com/downloads//).

[35] Tony Campbell, "A detailed reassessment of the Carte Pisane: a late and inferior copy or the lone survivor from the portolan charts' formative period?", *Map History*, retrieved January 2025, https://www.maphistory.info/CartePisaneConclusions.html.

also more accurate than contemporary maps, since they were based on navigational data collected by many pilots across many trips.[36]

Portolan charts were effective and accurate navigational aids, but there is one thing that becomes readily apparent when comparing them with a current map: the entire Mediterranean appears tilted nearly 10° eastward (Fig. 1.4). The sailors who had provided the navigational used uncorrected compasses which pointed to magnetic north and not true north (the heading to the North Pole). The difference between the two is called magnetic declination, and changes over space and time because the Earth's magnetic field is neither uniform nor static. It is unlikely that European sailors were concerned with magnetic declination because it was reasonably homogenous across the

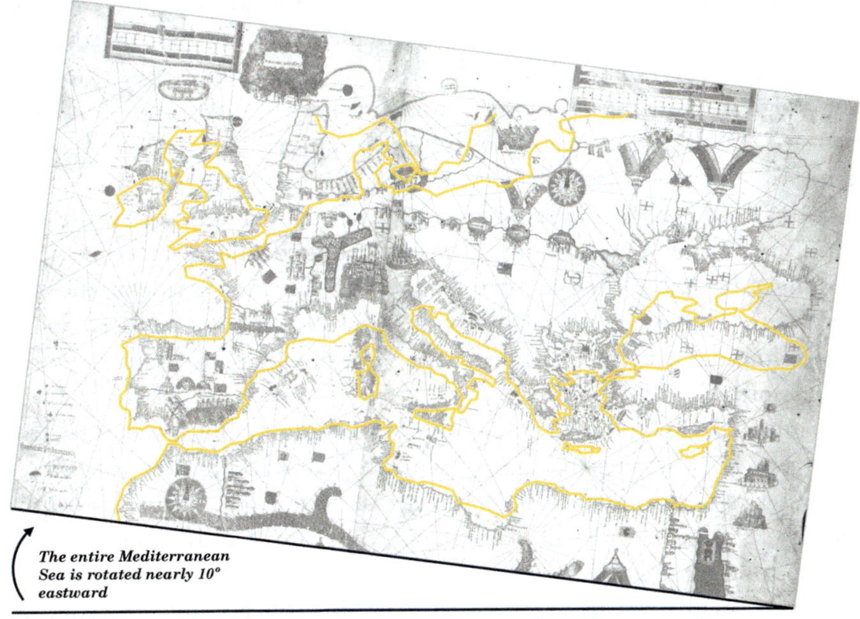

Fig. 1.4 A 1489 portolan chart made by Albino de Canepa overlaid with an accurate representation of the Mediterranean Sea, showing the nearly 10° eastward rotation caused by magnetic declination[37]

---

[36] Joaquim Alves Gaspar, "The origin of nautical cartography: certitudes, doubts, and perplexities", *International Journal of Cartography*, September, 2023, p. 1–30, on p. 14–16, doi:10.1080/23729333.2023.2240902.

[37] Albino de Canepa, *Chart of the Mediterranean, Black Sea and Western Europe* (Minneapolis: James Ford Bell Library, 1489mCa, 1489); accurate representation from continents adapted from Natural Earth […] using an equirectangular projection (WGS 84 / Plate Carrée ESPG:32662).

Mediterranean and did not change much across a century or so.[38] As long as they continued to navigate with the same uncorrected compasses which had been used to build the portolan charts, the issue was irrelevant. Conversely, if they decided to correct their compasses to point to true north, then the charts would send their ships some 10° off-course.

When Portugal started exploring the western coast of Africa in search of a maritime route to India, it quickly became obvious that the existing navigation techniques were not suitable for oceanic sailing. Dead reckoning (a way to estimate the ship's position based on travelled heading and distance) worked well within the limited confines of the Mediterranean, but estimation errors quickly compounded in offshore sailing, as there were no landmarks to double-check the ship's position. The solution was to bring aboard astrolabes or quadrants (already used by astronomers on land) to measure latitude, providing pilots with an absolute measurement of the ship's north–south position. Dead reckoning was now only required to estimate the ship's east–west progress, reducing positional errors and the risks of getting lost in the open sea.[39] This new navigation technique prompted the creation of latitude charts, which added a latitude scale onto the portolan charts.

At this point, magnetic declination became conspicuous, for two reasons: measuring latitude by astronomical means required finding true north; and the longer distances showed that magnetic declination was not the same everywhere. For instance, Columbus, in his 1492 Atlantic passage, claimed to have found a 'true meridian' near the Azores, where his compass pointed to true north.[40]

Magnetic declination had come out of the shadows, but it was still inconsequential. If pilots kept navigating with the same techniques (uncorrected compasses and astronomically determined latitude) that had been in place when the latitude charts had been drawn, then they had no issues retracing the steps of their precursors.[41] An analogy can be made with time: if the residents of a small town use the church clock for timekeeping, everyone will show up punctually for work and for dinner, even if the old clock runs fast in the summer and slow in the winter.

---

[38] This does not mean the phenomenon was unknown. For instance, Chinese measurements of magnetic declination date as early as the eighth century (Peter Smith and Joseph Needham, "Magnetic Declination in Mediaeval China", *Nature*, 214, p. 1213–1214, 1967, on p. 1213, doi:10.1038/2141213b0).

[39] See What Navigation Techniques Were Available to Magellan?

[40] Christopher Columbus, *Journal of the First Voyage of Columbus*, edited and translated by Clements R. Markham (London: Hakluyt Society, 1893), on p. 24, 25.

[41] At least for the same century for so. For longer periods the secular variations in magnetic declination required latitude charts to be updated.

There was however a price to pay for ignoring magnetic declination. Latitude charts were useful navigation tools, aiding pilots in finding their way to and from home, but they were bad maps, because they distorted the shape of continents. João de Castro, a nobleman that would become viceroy of Portuguese India, seems to have been the first to understand the issue.[42] In 1538, while sailing from Lisbon to Goa, Castro made systematic measurements of magnetic declination throughout the entire voyage. Noting that declination was positive (eastward) in the Atlantic and negative (westward) in the Indian Ocean, he correctly concluded this caused distortions and exaggerated the distance between Lisbon and Goa. Castro's reasoning reflected how latitude charts were made: much like the Portolan charts had been based on the reports of Mediterranean pilots, latitude charts leveraged the descriptions of the Portuguese explorers who had pioneered the maritime way to India. When running down the western coast of Africa, these pilots plotted the ship's position shifted to the east, because their uncorrected compasses were biased eastward (to starboard). After rounding the Cape of Good Hope, magnetic declination turned westward, but the pilots continued to shift their position to starboard, because they were now sailing upward. When cartographers turned the pilots' logbooks and descriptions of the coastline into charts, the resulting shape of Africa was elongated and shifted to the east (Fig. 1.5).

In 1517, when Magellan showed his chart to the Spanish King, Castro was only 17 years old, and the implications of magnetic declination had not yet exposed. While irrelevant for navigation, these implications were certainly not so for Magellan's claims, as the distortion of the African continent pushed Asia easterly, tipping the Moluccas into the Spanish hemisphere.

Latitude charts had other sources of error, of course. Ptolemy's geography—lingering but still influential—also overstated the length of Eurasia, navigators often exaggerated estimated distances (either to bolster their feats or as a safety margin, since it was better to spot land later than expected than to run into it in the middle of the night), and the lack of a projection created geometric inconsistencies.[43] However, magnetic declination seems to have been the key reason why the Moluccas appeared on the Spanish hemisphere.

If magnetic declination in the fifteenth century had been the same on both sides of Africa, the continent would still be distorted but not shifted to the east. If declination values had been swapped (westward along West Africa and eastward along East Africa), then the representation of the continent would

---

[42] Joaquim Gaspar and Henrique Leitão, "What is a nautical chart, really? Uncovering the geometry of early modern nautical charts", *Journal of Cultural Heritage*, 29, 2018, p. 130–136, on p. 134, doi:10.1016/j.culher.2017.09.008, on p. 134.

[43] See Why Did Nautical Charts Ignore the Earth Was Round?

Fig. 1.5 In 1538, João de Castro's sailed from Lisbon to Goa and measured magnetic declination along the way, correctly concluding it distorted charts and exaggerated the distance between Portugal and India[44]

shift westward. Both these scenarios would have likely left the Moluccas well within the Portuguese hemisphere. In other words, Magellan's expedition had been inspired by an arbitrary arrangement of the Earth's molten metallic core.

[44] João de Castro, *Roteiro de Lisboa a Goa (1538)* (Lisboa: Agência Geral das Colónias, 1940); accurate representation of continents adapted from Natural Earth [...] using an equirectangular projection (WGS 84 / Plate Carrée ESPG:32662); underlying chart adapted from *Kunstmann IV* [...].

2

A Rougher Atlantic Crossing

From Tenerife (October 3rd, 1519) to Rio de Janeiro (December 13th, 1519)

After taking in additional provisions and making a few last-minute changes to the crew, the fleet was ready to leave the Canary Islands and cross the Atlantic (Fig. 2.1). Before it left though, a caravel entered the harbour bearing an urgent message for Magellan. It was a letter from his father-in-law warning him that while the fleet was still at Sanlúcar de Barrameda some of his Spanish captains had boasted openly about removing Magellan from command, killing him if necessary[1].

He had heard similar threats before, but now the warnings came from his own father-in-law, and Magellan was bound to take them more seriously. After all, turning his back to Portugal to serve Spain was treacherous if not treasonous, so it would be folly to expect the Spaniards to blindly trust him.

Juan de Cartagena, the captain of the *San Antonio*, seemed the most likely to lead a mutiny. Cartagena was the nephew (or illegitimate son, as rumours would have it[2]) of Juan Rodríguez de Fonseca, a bishop closely involved in the preparation of the expedition and who ruled over the *Casa de Contratación*—the House of Trade, an agency to which the Crown delegated a broad range of matters related to overseas territories. When the king ordered all accounting and monetary affairs of the endeavour to be under the responsibility of a Spaniard, Fonseca used his influence to place his nephew as the fleet's chief

---

[1] José Toribio Medina, *El descubrimiento del Océano Pacífico: Vasco Núñez de Balboa, Fernando de Magallanes y sus compañeros* (Santiago de Chile: Imprenta Universitaria, 1920), on p.CXCIII, citing Correa, Barros, and Argensola.

[2] Laurence Bergreen, *Over the Edge of the World: Magellan's Terrifying Circumnavigation of the Globe* (New York: Harper Perennial, 2004), on p. 55.

R. Gaspar, *The Revolution of Magellan*, https://doi.org/10.1007/978-3-032-10797-8_2

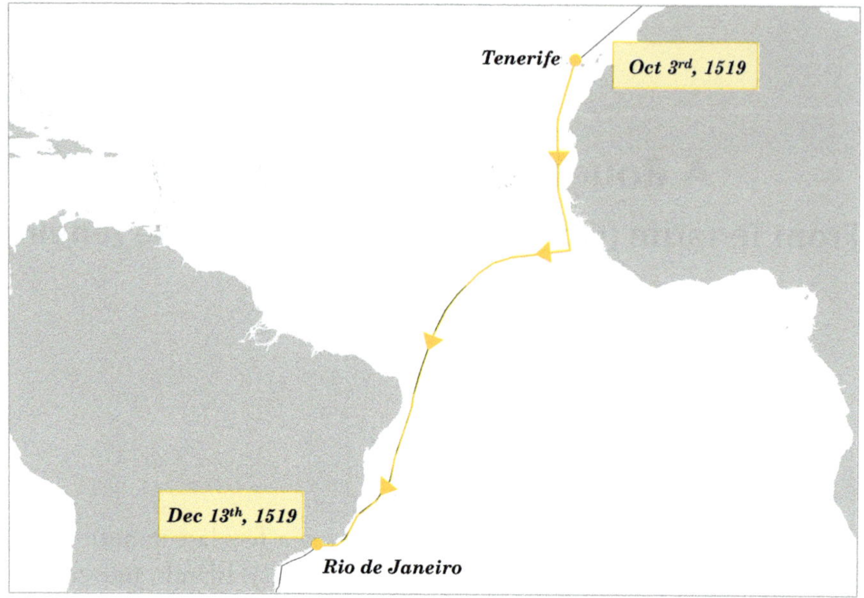

Fig. 2.1 Fleet route[3]

inspector and captain of its largest ship.[4] Since Faleiro, Magellan's initial partner, was no longer part of the expedition,[5] the king also determined that Cartagena would be Magellan's *conjunta persona*, or "joint person".[6] The term was somewhat vague but could fairly be construed as an order to treat Cartagena as a peer rather than a subordinate, to be consulted in all major decisions. His powerful role, together with his influential uncle, emboldened Cartagena to openly defy Magellan's leadership well before the voyage even started.

Gaspar de Quesada and Luis de Mendoza, respectively the captains of the *Concepcion* and of the *Victoria,* both Spanish, would also undoubtedly stand with Cartagena in a mutiny. Mendoza's contempt for Magellan seemed even more visceral than Cartagena's. Shortly after arriving to Seville, he had been

---

[3] Fleet route adapted from Tomás Mazón Serrano ("Mapas", Ruta Elcano, June 19, 2024, https://rut aelcano.com/mapas/); map adapted from Natural Earth (1:10 m large scale land data, version 5.1.1, https://www.naturalearthdata.com/downloads/).

[4] Tim Joyner, *Magellan* (Maine: International Marine, 1992), on p. 90, 91.

[5] See How Did Magellan Convince the Spanish King?

[6] Martín Fernández de Navarrete, *Coleccion de los viajes y descubrimientos que hicieron por mar los españoles desde fines del siglo XV* (Madrid: Imprenta Nacional, 1837), in Tomo IV, p.L.

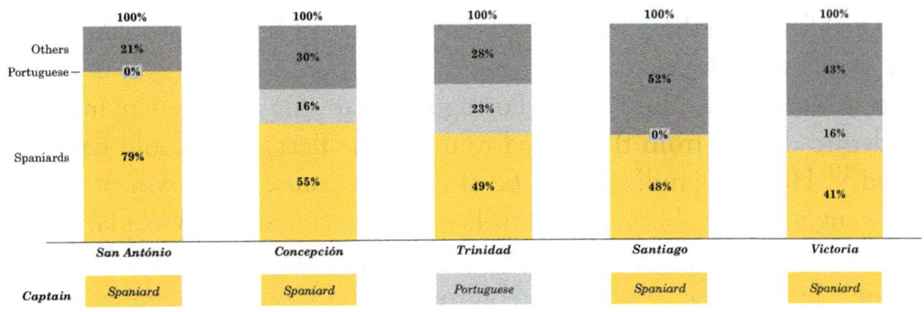

Fig. 2.2 Proportion of nationalities on each ship[8]

involved in a public row with Magellan which earned him a strong reprimand from the king.[7]

Juan Rodríguez Serrano, the captain of the *Santiago*, was also Spanish but appeared less interested in patriotic disputes. Serrano was an experienced mariner who had already crossed the Atlantic and served as a royal pilot. In the case of a mutiny, he would conceivably side with the faction which offered better prospects of successfully completing the mission. The same could be said of the many foreign professionals—Italians, French, Germans, Greeks—who made up a substantial part of the crew.

Most of the Portuguese mariners, many of them sailing on the *Trinidad*, would likely side with Magellan in a mutiny. All things considered though, Magellan's chances were precarious: his Portuguese faction represented just 12% of the mariners, compared to Cartagena's 56% Spanish majority. The remaining 32% of foreigners could swing either way (Fig. 2.2). Surely Magellan could not sit idly while the insurgency brewed.

What happened next has been interpreted over the centuries as either unabated arrogance or unmistakable evidence of Magellan's leadership skills. As often happens, the truth probably lies somewhere in the middle. Twelve hours after leaving Tenerife, Magellan's *Trinidad* flagship signalled a course change. Rather than continuing sailing southwest towards the Cape Verde islands, as had been agreed with the *Casa* and re-confirmed with the Spanish captains in Tenerife, the *Trinidad* steered the fleet closer to the African coastline. After a couple of days, when the *San Antonio* approached the *Trinidad* for the evening salute, Cartagena complained about the new course not being

---

[7] Jean Denucé, *Magellan: la question des Moluques et la première circumnavigation du globe* (Brussels: Hayez, 1911), on p. 240.

[8] Adapted from Tim Joyner's ship rosters, excluding unknown nationalities: *Magellan* (Maine: International Marine, 1992), on p. 252–263.

discussed first with the captains, and that the current heading would place the fleet too close to the coast.[9]

To avoid an encounter with Portuguese war ships, Magellan may have decided to deviate from the agreed route, as the fleet's plan could have been leaked.[10] He was familiar with the alternative course, as it was often used by Portuguese ships headed to Brazil.[11] He could have easily explained this to Cartagena, perhaps earning back a bit of trust from the suspicious Spanish captains. He chose instead a more confrontational approach, irritably replying to Cartagena that his role was to follow orders, not to ask questions.[12]

The flippant way in which Magellan had ignored the king's direct orders of treating Cartagena as a *conjunta persona* left the proud Spanish nobleman fuming, but the ships continued to make steadfast progress along the African coast. Fifteen days after the incident, while close to Sierra Leone, the fleet was hit by a storm and forced to bring down all sails to avoid tipping over the ships.[13] The storm subsided but so did the wind and, for the next twenty days, the ships barely made any progress.

Cartagena once again used the evening salute to voice his disdain for Magallen's leadership. Rather than having a member of his crew hail the customary *"God save you, captain general sir!"*, he had him shout a mere *"God save you, captain sir!"*. The omission of "general" likely intended to show Cartagena considered Magellan his equal, not his superior. Making things worse, for the next three days Cartagena forbade everyone aboard the *San Antonio* of rendering any sort of evening salute to the flagship.[14]

Magellan's position was perilous. The *San Antonio* was the fleet's largest ship and, together with the *Concepcion* and the *Victoria*, vastly overpowered the *Trinidad*'s artillery. Without knowing why Magellan had deviated from the planned route, the Spanish captains probably had good reason to think the captain general's reckless behaviour was endangering the fleet and that removing him from his post was justified.

When tension was at its peak and no gambler would put his money on Magellan, a rather unexpected event created an opportunity for the

---

[9] José Toribio Medina, *El descubrimiento del Océano Pacífico* […], on p.CXCIV, CXCV

[10] Portuguese agentes in Seville spared no expense to derail the expedition, see Why Was Magellan, a Portuguese Citizen, Serving Spain?

[11] José Malhão Pereira, "Fernão de Magalhães e a Sua Viagem no Pacífico. Antecedentes e Consequentes, *VII Simpósio de História Marítima* (Lisboa: Academia de Marinha, 2002), p.343–361, on p.343.

[12] José Toribio Medina, *El descubrimiento del Océano Pacífico* […], on p.CXCV, citing Herrera.

[13] Antonio Pigafetta in H.E.J. Stanley, *The first voyage round the world, by Magellan. Translated from the accounts of Pigafetta, and other contemporary writers* (London: Hakluyt Society, 1874), on p.41.

[14] Tim Joyner, *Magellan* […], on p.122.

commander. Antonio Salomone, the master of the *Victoria*, was caught sodomising a cabin boy—an act punishable by death. Magellan took the chance to summon all captains and pilots to the *Trinidad* for a court martial. There, Cartagena again questioned the captain general's choice of route. When all his enquiries were met with silence and a cold stare, he grew increasingly confrontational. Once Cartagena's remarks trespassed into clear insubordination, Magellan sprang from his seat, grabbed Cartagena by his shirt and declared he was under arrest. Cartagena pleaded to his countrymen to seize Magellan instead, openly inciting a mutiny. Caught off guard and wary of the armed guards Magellan had stationed around the *Trinidad*'s poop deck, no one made a move. Cartagena was unceremoniously put into the stocks of the main deck, normally reserved for the punishment of common sailors. The other captains appealed to Magellan that this was no way to treat a nobleman and asked if one of them was allowed to keep custody of the prisoner, under a solemn oath to return him at the first request.[15]

With Antonio de Coca, the fleet's accountant, now in charge of the *San Antonio*, the ships eventually left the doldrums and made solid westerly progress. On the morning of December 13, they arrived at Rio de Janeiro. This sheltered bay with plenty of resources to replenish the ship's provisions was known to the Portuguese since 1502 and was likely Magellan's first intended stop in South America. If it had not been for the route taken, the fleet would have risked landing north of Cape São Roque, from where it would take months of fighting strong head winds to move down the Brazilian coast towards Rio de Janeiro.[16]

Much had changed in just two and a half months. Magellan, starting from a fragile position, had succeeded in imprisoning his main rival and perhaps convincing some of his sailors only he had the knowledge necessary to guide the expedition to good port. Magellan had been decidedly haughty and undeniably lucky, but he also proved to be a better plotter than Cartagena. The Spaniard, in place of leveraging the superior firepower of the *San Antonio*, the *Concepcion,* and the *Victoria* to force the *Trinidad* into surrendering, chose to run his mouth in the only place Magellan fully controlled. He then incited a rebellion in front of numerous witnesses, something which would not bode well for him in a court martial.

---

[15] José Toribio Medina, *El descubrimiento del Océano Pacífico* […], on p. CXCVII, CXCVIII, citing López Recalde and Juan Sebastían Elcano.
[16] Tim Joyner, *Magellan* […], on p.123, 124.

2.1   Why Was Magellan, a Portuguese Citizen, Serving Spain?

The short answer

Magellan was a low-ranking but ambitious Portuguese nobleman. In 1505, he saw in the Indies campaigns an opportunity to prove his valour. After the better part of a decade and several injuries, Magellan returned to Lisbon expecting the king to recognise his contributions and valuable first-hand knowledge of the East. King Manuel I, however, saw in him nothing more than a soldier and sent him to Northern Africa to help quelch a rebellion. Magellan's proudness and lack of court savviness further distanced him from the king's good graces. In his last royal hearing, after having his request for a pension raise once more denied, he asked for permission to serve another king. Together with other individuals who also felt short-strawed by the Portuguese monarch, Magellan turned to Spain for an opportunity at glory and fortune. By the time Manuel I realised his mistake, the young Spanish king had already approved Magellan's proposal to reach the Moluccas by travelling west. Vasco da Gama suggested killing Magellan, but Manuel I did not want to jeopardise his country's diplomatic relations with Spain. Instead, he embarked on a wearing but ultimately unsuccessful attempt to sabotage the expedition.

Magellan was the son of a Portuguese *fidalgo*, the fourth of the five grades of nobility below royalty. The Magellans were descendants of a French crusader who had fought in the eleventh-century campaigns against the Moors, conquering the lands which would later form the Kingdom of Portugal. Centuries had eroded the family's standing, but they were still sufficiently close to King João II to get Magellan and his brother Diogo de Sousa accepted as pages of the queen.[17]

At the Lisbon court, Magellan had a better education that what was typical for lesser noblemen. Besides learning the usual court skills—reading, writing, arithmetic, horsemanship, combat—he could have been educated in mathematics, astronomy, and navigation. Magellan conceivably also had access to the *Casa da Índia* (House of India)—the equivalent to Spain's *Casa de Contratación*—and to the Council of Mathematicians—a gathering of the country's leading cosmographers and astronomers—which could have advanced his knowledge of the practical and theoretical aspects of navigation.[18]

---

[17] Diego Barros Arana, *Vida y Viages de Hernando de Magallanes* (Santiago de Chile: Imprenta Nacional, 1864), on p.18, citing Argensola.

[18] Tim Joyner, *Magellan* (Maine: International Marine, 1992), on p.36.

It was also through his education that Magellan met the future King Manuel I. At the time, Duke Manuel—cousin and brother-in-law to King João II—was responsible for the education of the court's pages. He was also at odds with his cousin. In 1491, João II's legitimate son had died in a mysterious fall, leaving Manuel as heir to the throne. However, João II had been trying to position his illegitimate son Jorge as his heir, estranging Manuel and half of the court in the process.[19] When King João II died in 1494, Manuel, rather than Jorge, was crowned king, which left Magellan and his brother in a difficult position. Their family had been loyal to the old king and was now on the bad side of the new ruler.

Magellan's next years did not seem to amount to much. It was only in 1505, when already about twenty-five years old, that he managed to set foot on a ship. Magellan, his brother, and his soon-to-be close friend Francisco Serrão enrolled as supernumeraries[20] in a crew of thousands sailing with Francisco de Almeida—the future first viceroy of Portuguese India—to seize control of shipping routes in the Indies.

Magellan spent the next seven years or so in the Indies, amassing first-hand experience he would later use to persuade the Spanish king to finance his expedition. Official records rarely mention him—commanders and court chroniclers did not waste much ink on subordinates—but, when Magellan's name does come up, it is usually to highlight episodes of valour or to list the injuries that often accompanied them (Fig. 2.3).

One of the first written references of Magellan is indeed about an injury he sustained while serving the Portuguese Crown on the naval battle of Kannur (south of Goa), in the fall of 1505. A word of praise is also reserved for Serrão, for bravery beyond the call of duty.[21] In 1506 Magellan was back at the other side of the Indian Ocean, in Kilwa (modern-day Tanzania), where he was awarded the captainship of an oar-propelled brigantine after earning the respect of his commanding officer.[22] This was seemingly the only time Magellan commanded a ship, but he continued to show up episodically in mission reports. In 1509, he was wounded at the naval battle of

---

[19] H. V. Livermore, *A New History of Portugal* (Cambridge: Cambridge University Press, 1966), on p.132.

[20] Additional members of the crew not required for the day-to-day operation of the ship, but expected to defend the fleet and support the crew when needed (see What Was the Life of a 16th Century Sailor Like?)

[21] Jean Denucé, *Magellan: la question des Moluques et la première circumnavigation du globe* (Brussels: Hayez, 1911), on p.102.

[22] Visconde de Lagôa, *Fernão de Magalhães: A sua vida e a sua viagem* (Lisboa, Seara Nova, 1938), in Volume I, p.127.

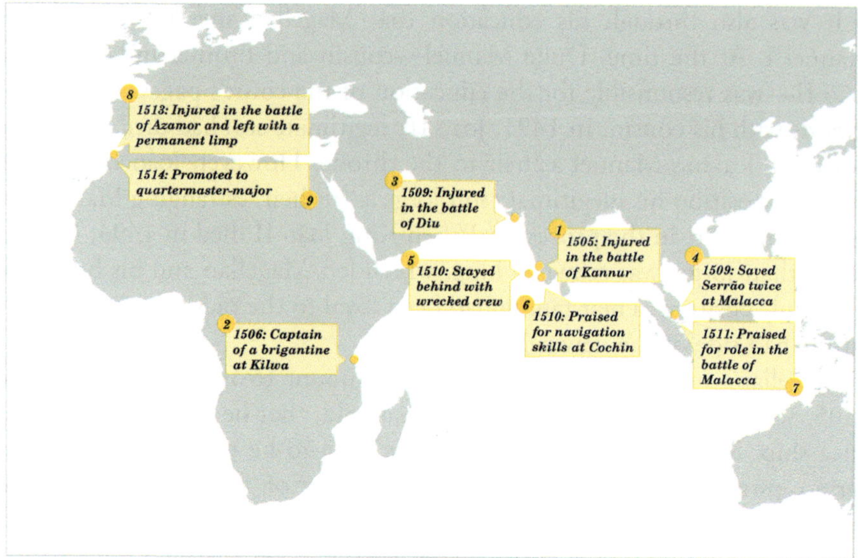

Fig. 2.3 Main court records about Magellan's career in the Indies

Diu, while fighting hand-to-hand with the crew of an Egyptian warship[23]. A few months later he saved his friend Serrão twice in a row, first from an ambush at Malacca[24] and shortly after from a failed attempt to seize a junk.[25] Magellan seemed to have a penchant for rescue missions—and his superiors a fixation for raiding junks—as not long after he found himself amidst another botched pirating attempt. The Portuguese captain wanted to abandon his boarding party on the seized junk—sinking rapidly after its crew succeeded in breaching the hull—but Magellan volunteered to retrieve them.[26] The following year, in 1510, a ship taking Magellan grounded near the Laccadive Islands, off the coast of India. While the officers and nobles clambered aboard the few support vessels, Magellan stayed behind with the crew to reassure them a rescue ship would be sent (not at all a given in the sixteenth century).[27] The same year, in Cochin (southern India), Magellan's good understanding of navigation matters was noted in a council convened

---

[23] Gaspar Correa, *Lendas da Índia* (Lisbon: Typografia da Academia Real das Ciências, 1860), in Tomo II, p.28.

[24] José Toribio Medina, *El descubrimiento del Océano Pacífico: Vasco Núñez de Balboa, Fernando de Magallanes y sus compañeros* (Santiago de Chile: Imprenta Universitaria, 1920), on p.XXI, citing Barros.

[25] Visconde de Lagôa, *Fernão de Magalhães* [...], on p.131, citing Castanheda and Góis.

[26] Idem, on p.132.

[27] Jean Denucé, *Magellan* [...], on p.109, citing Herrera.

by Afonso de Albuquerque,[28] Portuguese India's second viceroy. Magellan's name surfaced again in 1511, where he provided *"a very good account of himself"* when Portugal took control of Malacca.[29]

Afterwards, Magellan and Serrão, already bound by a brotherly friendship, split up. Serrão was one of the captains in a fleet tasked with finding the Moluccas, while Magellan seems to have stayed behind in Malacca.[30] The fleet never made it to the Moluccas, as it acquired large quantities of nutmeg and mace halfway through the trip and returned to Malacca to store them. On the way back, a storm separated the ships, and the vessel carrying Serrão grounded in an islet, stranding him and about ten companions.[31] After seizing a junk which had come to investigate the wreck, the troupe resumed the expedition's mission, finding their way to the Moluccas and becoming the first Portuguese to set foot there. Instead of returning to Malacca, Serrão stayed as the military advisor for one of the local sultans, sending letters to his superiors and to Magellan with enticing descriptions of the wealth of the islands and the opportunities they offered.[32]

Sometime in 1512 or 1513, Magellan returned to Lisbon with the expectation of bright days ahead. Besides some seven years of faithful service to the Crown under his belt, he now had a trusted friend living in the Moluccas, the source of clove which had remained a secret to Europeans for centuries. He also expected to have some money waiting for him at Lisbon. A few years before, Magellan had lent most of his meagre savings to Pedro Abraldez, a Portuguese merchant who needed capital to buy pepper in the Indies and bring it back to Portugal. However, Abraldez had died and his father fled Portugal to escape his son's many creditors.[33]

Magellan's finances were now even worse than before, but certainly the king would recognise his deeds and injuries as proof of loyalty and dedication, and reward him accordingly? While spice trade had turned King Manuel I into one of the richest monarchs in Europe, he was not known for being generous. Worse, the king's attitude towards Magellan was at best

---

[28] Idem, on p.110, citing Afonso de Albuquerque.

[29] Idem, on p.115, citing Herrera.

[30] Visconde de Lagôa, *Fernão de Magalhães* [...], on p.147. There are authors who argue Magellan also travelled through Southeast Asia and visited the Banda Islands (just south of the Moluccas), but that is not the current consensus. For more information on that alternative hypothesis see José Manuel Garcia, "Fernão de Magalhães o homem que descobriu o mundo tal como ele é", *Revista Mátria XXI*, 2021, p.313–327, on p.314–326.

[31] Idem, on p.122.

[32] Jean Denucé, *Magellan* [...], on p.124, 125; José Toribio Medina, *El descubrimiento del Océano Pacífico* [...], on p.XXVII.

[33] Visconde de Lagôa, *Fernão de Magalhães* [...], on p.148.

indifferent. Instead of rewarding his endeavours and recognising his navigation skills and first-hand knowledge of the Indies, the Crown treated him like any other soldier. Tother with many others, Magellan was shipped over to the Portuguese territories in Azamor (modern-day Morocco), to quelch a rebellion. He was assigned to a cavalry unit and required to buy his own horse. The string of offences did not seem to curb Magellan's willingness to put himself in peril, as he was a regular on the front line and on cavalry raids. In one of those excursions a lance injured his knee, leaving him with a permanent limp. A crippled man was of little use on the front line, but his commanding officer saw potential in Magellan and promoted him to the coveted post of quartermaster major.[34]

It was a big step-up for Magellan, but it would not amount to more than a brief hump in an otherwise steep downhill. In a succession of miscalculated steps, Magellan quickly turned the king's indifference for him into barely disguised irritation. Before being promoted to quartermaster, Magellan's horse was killed in combat. The horse was probably not the finest equine specimen, as he was offered less than 30% of the going rate for this sort of loss. Rather than bringing the issue to the attention of his immediate superiors, Magellan sent a letter directly to the king.[35] Somebody less stubborn and more politically savvy would probably have realised Manuel I—a monarch who enjoyed the meanderings of court protocol—would be offended by the presumption that his direct intervention in such a petty matter was justified. For Magellan though, no grievance was small enough.

The next misstep came did not take long and was far more serious. As a quartermaster, Magellan was asked to distribute among the victors a large herd abandoned by the Moors—with some 200,000 goats and 3,000 camels and horses. Before the herd was corralled and inventoried though, Magellan gave 400 goats directly to the tribesmen that had fought alongside the Portuguese. A group of senior officers—who perhaps resented Magellan for overtaking them to the post of quartermaster—accused him of selling the goats to the enemy and pocketing the proceeds.[36] It was a serious accusation, as it involved not only misappropriating Crown property but also abetting the enemy. Logging a formal response with his commanding officer would have been the most sensible option, but Magellan decided to abandon his post without authorisation and return to Lisbon to demand an audience with the king.[37]

---

[34] Jean Denucé, *Magellan* […], on p.130, 131.

[35] Visconde de Lagôa, *Fernão de Magalhães* […], on p.149.

[36] Idem, on p.151.

[37] Jean Denucé, *Magellan* […], on p.132, citing Barros.

Perhaps expecting a good explanation for his return, the king granted him an audience. Rather than defending himself, Magellan asked for a pension raise to cover his sacrifices to the Crown. An astonished Manuel I promptly refused the raise and produced the letter he received from Magellan's commanding officer, complaining he had deserted his post with serious charges looming over his head.[38]

A frustrated Magellan returned to northern Africa to face the charges, which he managed to dismiss. After doing so he returned to Lisbon, this time with the proper licence, and requested another audience. The king refused to even look at the dismissal notice and once more denied the pension increase. Magellan then asked if there was any way he could further serve the Crown, to which the monarch replied there was not. Lastly, Magellan inquired if he could serve another king. The answer, which sent Magellan's dignity plummeting down but also opened a door to new opportunities, was delivered with undisguised disgust: yes, he could go wherever he pleased.[39]

In hindsight, Magellan's request to serve another country should have raised alarm bells in the Portuguese court. Having first-hand experience in the Indies and being a close friend of Serrão, Magellan held sensitive information about the Moluccas. The king and his advisors were well also aware of the growing discussion on whether the islands laid west or east of the demarcation line, and that Spain was preparing an expedition to look for a western passage in South America.[40]

Portugal's actions played directly into the hands of the neighbouring kingdom in more ways than one, as Magellan was not the only one disgruntled with his country. Serrão had urged his superiors to take advantage of his close relationship with a local sultan to strengthen Portugal's position in the Moluccas, but was ignored.[41] Faleiro, Magellan's original partner in the expedition, held a grudge against the king for not adequately rewarding his contributions to the advancement of astronomy.[42] Cristóbal de Haro, one of the expedition's main financiers, had also been driven away from Portugal by Manuel I. After financing several of the country's maritime expeditions, Haro had lost seven ships off the Gulf of Guinea to a corsair attack led by Estevão Yusarte, a Portuguese privateer. King Manuel I had ordered the arrest

---

38 José Toribio Medina, *El descubrimiento del Océano Pacífico* [...], on p.XXXIV.

39 Jean Denucé, *Magellan* [...], on p.137.

40 See How Did Magellan Convince the Spanish King?

41 Visconde de Lagôa, *Fernão de Magalhães* [...], on p.156.

42 José Toribio Medina, *El descubrimiento del Océano Pacífico* [...], on p.LXXXIV.

and punishment of Yusarte but refused to compensate Haro for the loss of the ships.[43]

In 1517, Manuel I finally realised he had underestimated Magellan. The rude and obnoxious little nuisance had, against the monarch's wildest expectations, been granted an audience with the young King Charles I of Spain (who would, in 1519, also be coronated as Charles V, the Holy Roman Emperor). His agents also informed him that Haro and Diego Barbosa (a Portuguese expatriate living in Seville, soon to become Magellan's father-in-law) had been rallying up support for Magellan and Faleiro's plan among the Spanish king's advisors.[44]

Manuel I was trying to arrange his marriage to Leonor[45]—the Spanish king's sister—and could not afford to start on the wrong foot with his future brother-in-law. His best option was to work in the shadows and try to quietly kill the endeavour without formally getting involved. Álvaro da Costa, Portugal's ambassador in Spain, was tasked with the covert operations. Unable to prevent the royal audience, Costa complained to Cardinal Adrian—one of the king's three key counsellors—that Magellan's expedition could damage diplomatic relations with Portugal. The cardinal was sensitive to this line of reasoning but had also been advised by Bishop Fonseca—to whom he delegated most maritime affairs, including the control of the *Casa*—that there could be merit to Magellan's claim the Moluccas fell on the Spanish hemisphere. The other two advisors, uninterested in maritime exploration but sensible to commercial opportunities, had heard from Haro there was money to be made in the trade of cloves.[46]

To the ambassador's chagrin, Charles I approved Magellan's expedition in less than a month.[47] The young king, barely 18 years old and born in Flanders, had moved to Spain just a year before. He was aware of the Treaty of Tordesillas, but the uncertainty surrounding the location of the Moluccas meant that pursuing the opportunity was not a direct affront to his brother-in-law. Unlike Manuel I, he had no previous dealings with Magellan, and perhaps took his curt and matter-of-fact manners, short and dark-skinned appearance, and noticeable limp as evidence of the navigator's experience and fortitude.

---

[43] Tim Joyner, *Magellan* [...], on p.68.

[44] Idem, on p.74.

[45] Demetrio Ramos Pérez, "Magallanes em Valladolid: La Capitulación", *Actas do II Colóquio Luso-Espanhol de História Ultramarina, 1975*, p.179–241, on p.216.

[46] Tim Joyner, *Magellan* [...], on p.79, 80.

[47] José Toribio Medina, *El descubrimiento del Océano Pacífico* [...], on p.CXIII.

The ambassador hurriedly wrote to Manuel I, advising him to lure Magellan back to Portugal, a man he considered to be *"of great spirit, very skilled in matters concerning the sea"*.[48] The Portuguese king and most of his council were against bringing Magellan back. Killing him was discussed as an alternative, with Vasco da Gama openly criticising the king for not having done so already. Wary of the implications it could have for his marriage, the king refused to have Magellan done away with.[49] Still, a whisper of these discussions probably found its way to Spain, as Magellan and Faleiro gained an armed escort.[50]

Costa doubled down on his efforts to sabotage the expedition. His attempts to make Magellan change his mind—by accusing him of treason and of tarnishing the good name of his family in Portugal—were unsuccessful, and so were his endeavours to send his father-in-law back to Portugal.[51] The ambassador then turned his attention to undermining Magellan's credibility among the expedition's key stakeholders. Likely in conjunction with some well-placed bribes, he played at the already tense strings of Spanish patriotism by accusing Magellan of being an untrustworthy and dishonourable turncoat who only had his own interests at heart. While ultimately unable to sabotage the expedition, Costa's actions significantly contributed to the entropy and conflict that engulfed its preparation.

Young King Charles I continued to support the expedition, but he and his advisors seemed now less sure of Magellan's intentions. Cartagena was named inspector general, to succeed in command to Magellan and Faleiro in case they died, and to ensure *"the said Portuguese follow the prescribed route"*. To safeguard the Crown's economic interests, all the fleet's fiscal officers would need to be Spaniards.[52]

In Seville, the Portuguese ambassador tasked his local consul Sebastião Álvares with throwing the preparation of the expedition into havoc. Álvares apparently did so by distributing some bribes around the *Casa* and urging the crew not to sail under the command of a Portuguese traitor. In one instance, an agent of Álvares nearly succeeded in getting an enraged mob to raid the *Trinidad*. That day, the fleet's flagship was not flying the Spanish royal insignia at the masthead—which had been taken down for redyeing—but only Magellan's smaller family crest banners. The agent brought the crowd to

---

48 Visconde de Lagôa, *Fernão de Magalhães* [...], on p.210, citing Góis.

49 Jean Denucé, *Magellan* [...], on p.192, 193.

50 José Toribio Medina, *El descubrimiento del Océano Pacífico* [...], on p.CXXXV, CXXXVI, citing Las Casas.

51 Idem, on p.CXXXIII, CXXXIV.

52 José Toribio Medina, *El descubrimiento del Océano Pacífico* [...], on p. CXXIX.

an uproar by convincing them the family banners were Portuguese insignias, and Magellan made matters worse by refusing to take them down. The situation was only deescalated by Sancho de Matienzo, the *Casa's* treasurer.[53] Magellan reacted to these incidents as he always did, by complaining to the king. Unlike Manuel I, Charles I was receptive to this approach and intervened in his favour, ordering the offenders to be punished and Magellan to be elevated to the rank of Commander of the Order of Santiago.[54]

Towards the end of the preparations and with his coronation to Holy Roman Emperor quickly approaching, the king grew more distant from the expedition, leaving most matters in the hands of Bishop Fonseca and the *Casa*. Without the protection of the king, Magellan faced increasing hardships and delays, but eventually succeeded in getting the expedition underway, nearly three years after leaving Portugal.[55]

[53] Martín Fernández de Navarrete, *Coleccion de los viajes* [...] in Tomo IV, p.XLIV.
[54] José Toribio Medina, *El descubrimiento del Océano Pacífico* [...], on p. CXLIII, CXLIV, CXXXVI.
[55] Tim Joyner, *Magellan* [...]on, p. 97–113.

3

Out of Portuguese Territories

From Rio de Janeiro (December 26th, 1519) to the Tordesillas Meridian (December 30th, 1519)

By stopping at Rio de Janeiro, Magellan openly disregarded his orders to steer clear of Portuguese territories, something the king had insisted upon to avoid a diplomatic incident with Portugal. From Magellan's perspective, the benefits outweighed the risks. After the eventful Atlantic crossing, his crew could use some rest and fresh provisions (Fig. 3.1). He was also aware the Portuguese had not yet established a permanent settlement at Rio, so the likelihood of coming face to face with them was relatively low. Finally, he knew the indigenous there were friendly. This last point was not to be taken for granted, as a Solis had learned the hard way in 1516.[1]

Much about what Magellan knew about Rio de Janeiro was rooted on first-hand experience from João Lopes Carvalho, the Portuguese pilot of the *Concepción*. Carvalho, before following Magellan to Seville, had spent four years in Rio overseeing the storage facility of a Portuguese trader. During that time he become quite familiar with the locals, perhaps a little too familiar. Shortly after the fleet arrived, a woman with a seven-year-old boy approached Carvalho, claiming he was his son. The Portuguese seemingly had no qualms acknowledging the child as kin, and enlisted him as his page. The boy became known as Joãozito—"Little João", after his father—and would be the first Brazilian to cross the Pacific Ocean.[2]

The locals, likely Tamoios, welcomed the crew as warmly as they had welcomed Carvalho. The five ships were swarmed by canoes loaded with

---

[1] Solis was reportedly eaten by natives, see How Did Magellan Convince the Spanish King?

[2] Antonio Pigafetta, *The first voyage around the world 1519–1522*, edited by Theodore Cachey Jr. (Toronto: University of Toronto Press, 2007), on p. 137, 138.

R. Gaspar, *The Revolution of Magellan*, https://doi.org/10.1007/978-3-032-10797-8_3

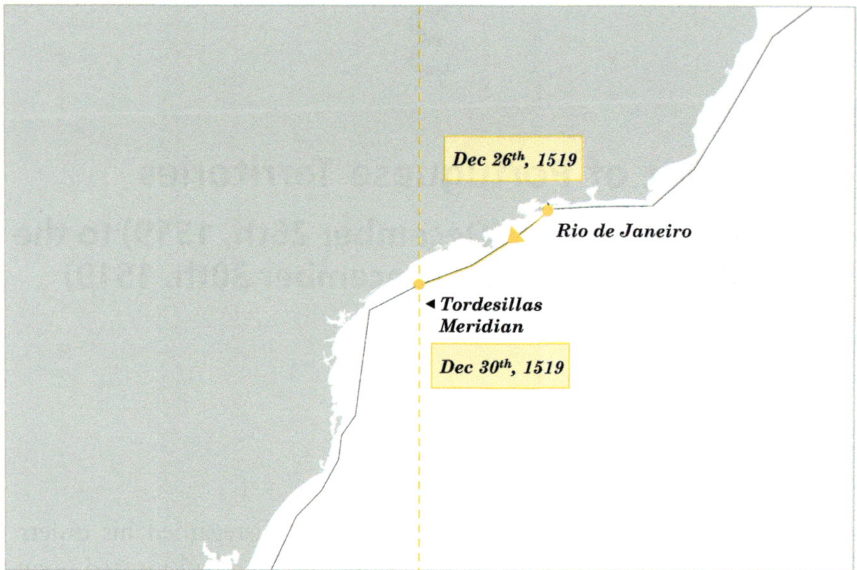

Fig. 3.1 Fleet route[3]

produce and animals to trade. Stark naked, devoid of body hair and adorned with the plumages of parrots, they did not go unnoticed by the crew.[4] Magellan reminded the wide-eyed sailors the residents should be treated with respect and women should not be molested. This rule was part of a set of 74 fleet orders, decreed by the king in detail and handed over to Magellan a few months before the expedition had left Spain.[5] As the expedition progressed, respect for the natives and for the king's orders would rapidly take a turn to the worse, but at Rio things remained somewhat civil.

Much of what is known about those days at Rio comes from Antonio Pigafetta's account of the journey. A descendent of a wealthy and influential Italian family, he joined Magellan's expedition as a supernumerary, hoping to see the world. Pigafetta kept a diary—which he would later turn into a book[6]—but it was only at Rio that his pen truly came alive, enthusiastically describing the local culture, fauna, and flora.

---

[3] Fleet route adapted from Tomás Mazón Serrano ("Mapas", Ruta Elcano, June 19, 2024, https://rutaelcano.com/mapas/); map adapted from Natural Earth (1: 10 m large scale land data, version 5.1.1, https://www.naturalearthdata.com/downloads/).

[4] Idem, on p.9.

[5] "Instrucción dada por el Emperador a Fernando de Magallanes y Ruy Falero [...]" in Cristóbal Bernal, *Crónicas de la Primera Vuelta al Mundo, según sus Protagonistas*, 2016, on p. 142–172.

[6] See How Do We Know What Happened in the Expedition?

He first directed his astonishment to the trades being done. The indigenous did not possess the knowledge to produce iron, and so would happily exchange five or six birds for a fishhook, or enough fish for ten men for a pair of scissors. Some trades went well beyond produce, with fathers willing to letting go of their daughters for a hatchet or a large knife. Pigafetta was particularly impressed with his own negotiation skills after swapping a playing card for five fowls. Pigafetta also enthusiastically described the delectable local food, from pineapples to sweet potatoes. Alas, not everything was to his liking: *"They make round white bread from the marrow substance of trees, which is not very good"*.[7]

The Europeans also made quite the impression on the locals. Shortly after the fleet arrived it started raining, ending a two-month drought and convincing the natives Magellan and his crew had something to do with it. The two masses held ashore further reinforced their belief the Europeans commanded the skies. When seeing the hardened mariners kneeled over a cross, the locals did the same and piously raised their clasped hands in the air.[8]

However, the Tamojos also got to witness a far darker side of the visitors. Antonio Salomone, the *Victoria's* master caught engaging in sodomy with an apprentice during the Atlantic crossing, was sentenced to death and garrotted in public.[9] Antonio Genovés, the cabin boy in question, was spared due to his young age but had to endure the mockery and ridicule of his shipmates. Four months later he drowned after throwing himself overboard.[10]

Magellan too was forced back to his dark place by hints of a mutiny brewing. Cartagena, purportedly in custody, had managed to go ashore with the help of Antonio de Coca, who had replaced him as the captain of the *San Antonio*. The two Spanish noblemen argued they were simply enjoying the time off, but Magellan suspected they were rounding up supporters for a mutiny. Both were arrested and the *San Antonio* received its third captain, Álvaro de Mesquita, a Portuguese who had so far served as a supernumerary on the *Trinidad*. Magellan threatened to maroon Cartagena when the fleet

---

[7] Antonio Pigafetta, *The first voyage around the world 1519–1522* [...], on p. 10.

[8] Idem.

[9] José Toribio Medina, *Colección de documentos inéditos para la historia de Chile: desde el viaje de Magallanes hasta la batalla de Maipo: 1518–1818* (Santiago de Chile: Imprenta Ercilla, 1888), in Tomo I, p. 171.

[10] *1ª Declaración (actualizada) de las personas fallecidas en el viaje al Maluco (del 20-XII-1519 al 29-VII-1522)* in Cristóbal Bernal, *Crónicas de la Primera Vuelta al Mundo, según sus Protagonistas*, 2016, on p. 243.

departed but, once again, acceded to the pleas of lenience of the Spanish captains.[11] Both Magellan and Cartagena would soon regret that decision.

After thirteen days in Rio de Janeiro, the fleet left amidst of chorus of complaints from the crew. Not wanting to push his luck and eager to search for a passage to the Indies, Magellan gave the departure order shortly after Christmas Eve. A few days later, the fleet crossed the Tordesillas demarcation line, at a longitude of about 47°. However, his nautical charts showed the South American coastline shifted to the east,[12] so Magellan likely thought he was still in Portugal's hemisphere, prompting him to continue racing down the coastline.

3.1   What Navigation Techniques Were Available to Magellan?

The short answer

Magellan's fleet sailed from Spain a mere 30 years after Columbus', but the navigation techniques of Magellan's pilots were already leaps and bounds ahead of those of the Italian admiral. The unreliable dead reckoning technique had been replaced by latitude navigation, which used astronomical measurements to determine the ship's north–south position. Pilots were no longer illiterate craftsman, but were instead required to know how to use quadrants and astrolabes, and how to read astronomical tables. Without these techniques, the expedition would not have been possible.

In 1492, Columbus made his first Atlantic crossing using only dead reckoning,[13] a way to track progress from heading (taken with a compass) and distance travelled (estimated from the ship's speed). The method worked well in the Mediterranean, where sailors could use known land markers to frequently double-check the ship's position, but was woefully inadequate for open-sea navigation. Under worse conditions, Columbus' voyage could have easily ended badly.

Less than three decades later, a revolution had taken hold of ocean exploration, prompted by the advances made by the Portuguese while trailblazing

---

[11] José Toribio Medina, *El descubrimiento del Océano Pacífico: Vasco Núñez de Balboa, Fernando de Magallanes y sus compañeros* (Santiago de Chile: Imprenta Universitaria, 1920), on p. CCIII.

[12] Due to magnetic declination, see How Did Magellan Convince the Spanish King?

[13] Samuel Eliot Morison, *Admiral of the Sea: A life of Christopher Columbus* (Boston: Northeastern University Press, 1983), on p. 183–196.

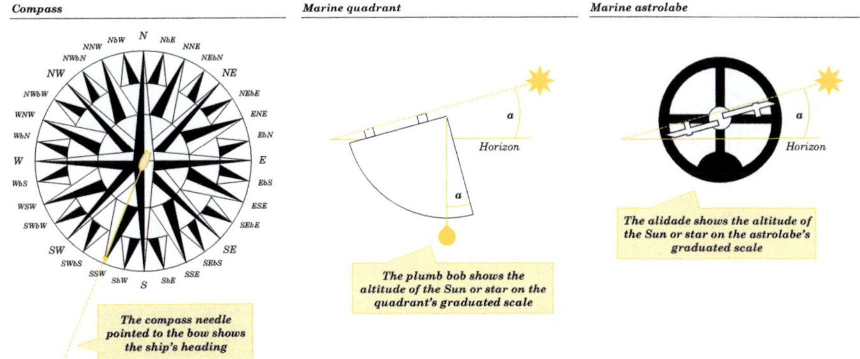

Fig. 3.2 Schematic of how a compass was used to measure the ship's heading, and marine astrolabes and quadrants to measure the altitude of the Sun and other stars

the maritime passage to India. All of Magellan's pilots eschewed dead reckoning in favour of latitude navigation, which allowed them to have an absolute fix on the ship's north–south position, even in the middle of the sea and far away from familiar coastlines.

However, taking a latitude reading was not as simple as looking at a compass; it required astronomical measurements and calculations. Magellan's pilots, much as their Portuguese peers who had negotiated their way around the Cape of Good Hope, had more reading and mathematical skills than the Mediterranean merchant pilots, but there were still limits to the complexity of the instruments and the extent of computations they could perform aboard a swaying ship. The solution was to simplify existing astronomical instruments, plus to develop easy-to-follow procedures based on step-by-step instructions, diagrams, and calculation aiding tables.[14] The toolset of these pilots thus kept the compass from the old dead reckoning ways, and added a marine quadrant and an astrolabe (Fig. 3.2).

Land astronomers used quadrants to measure angles at least since the days of Ptolemy. The marine quadrant was stripped of all superfluous features and anything that could easily be damaged aboard a ship, leaving only a wooden quarter circle with a plumb bob suspended from the vertex. To use it, the pilot aligned the instrument's two sights with the North Star and recorded the angle of the plum bob on a graduated scale. In the sixteenth century, the North Star was not exactly over the pole but described a small circle of

---

[14] Joaquim Gaspar, "Navegação e cartografia náutica nos séculos XV e XVI", *Ciência, tecnologia e medicina na construção de Portugal. Novos horizontes Sécs. XV-XVI* (Lisboa: Tinta da China, 2021), p. 99–124.

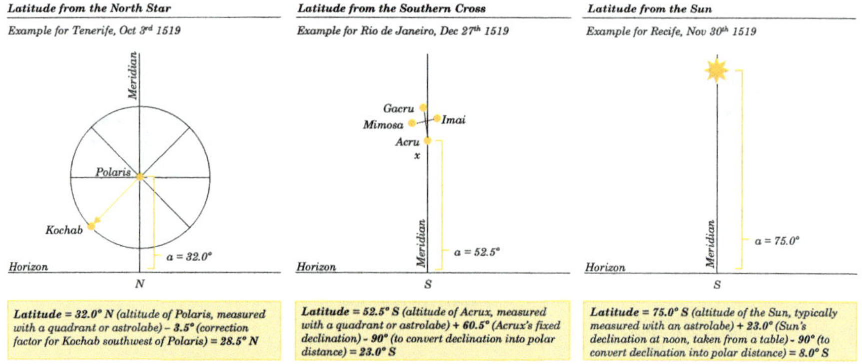

Fig. 3.3 Examples of latitude measurements using Polaris (the North Star), the Southern Cross constellation, and the Sun[16]

3.4 degrees around it.[15] To correct for this, the pilot compared the relative position of the North Star with that of Kochab, part of the Ursa Minor constellation. He would then use the "Procedure of the North" to know which correction factor to apply to the quadrant measurement to calculate the latitude of the place (Fig. 3.3).

Once explorers got near the equator, the North Star dipped below the horizon and could no longer be used to measure latitude. As there was no star close to the South Pole (Polaris Australis, the closest, is barely visible to the naked eye), the alternative was to measure the altitude of Acrux (the lower star of the easily identifiable Southern Cross constellation) at culmination.[17]

Latitude could also be obtained from the position of the Sun at solar noon (i.e., at culmination), measured with a marine astrolabe. Marine astrolabes were based on planispheric astrolabes (used in astronomy at least since ancient Greece) but stripped to its essentials: a graduated brass circle, skeletonised except for the bottom so it would not swing or spin as easily; and a pivoting alidade. The pilot would position the alidade so the sun shined through one of its sights and landed on the other one, and then read the angle from the graduated scale. Finally, he would turn to the "Procedure of the Sun" to convert the measured angle into latitude (see example on Fig. 3.3).

---

[15] Currently the difference has narrowed to 0.7 degrees, about one-and-a-half times the width of a full moon. It will be nearest to the North Pole in year 2100, before drifting away again.

[16] Adapted from Abel Fontoura da Costa, *A Marinharia dos Descobrimentos*, Lisboa, Agência Geral das Colónias, 1938, p. 48–143); examples of sixteenth century celestial positions based on the NASA JPL DE431 ephemerides (William Folkner et al., "The Planetary and Lunar Ephemerides DE430 and DE431", *IPN Progress Report 42–196*, 2014).

[17] At culmination a star is at its highest altitude in the sky and points directly south (or north, if the observer is in the Northern Hemisphere).

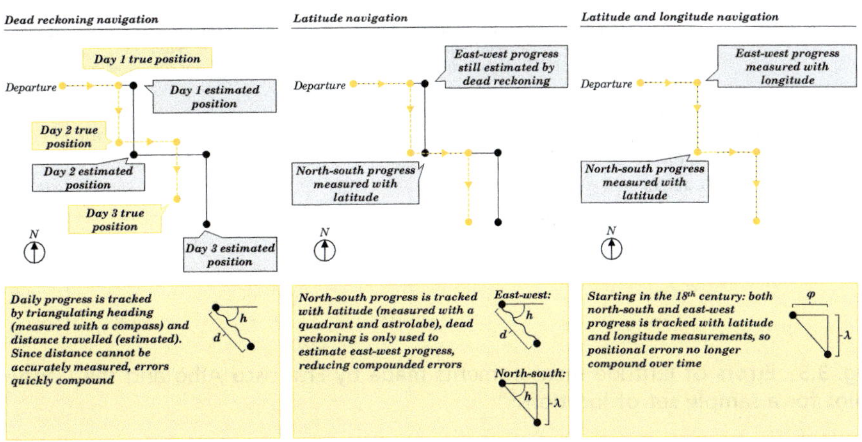

Fig. 3.4 Comparison of accumulated errors in dead reckoning, latitude navigation, and latitude/longitude navigation

Latitude navigation was a major improvement over dead reckoning, but it still only solved half of the problem. Without the means to measure longitude,[18] east-west headway had to be indirectly calculated from latitude and heading.[19] Under ideal conditions—say, to estimate progress from a single day of sailing with constant heading—this method was reasonably accurate. However, errors in the east-west position compounded over time, since each new estimate was based on the previous one (Fig. 3.4).

Sixteenth century pilots mitigated this issue in two ways. One was to use estimated distance as an additional piece of navigational information. Another is known today as *parallel sailing* or *running down the latitude*. For instance, pilots travelling from Lisbon to Madeira would, wind permitting, first sail south until they reached the latitude of the archipelago and then go west until they spotted land. The resulting voyage was longer than the more direct southwestern route but reduced the risk of missing the small islands. Experienced pilots were also able to infer they were approaching land by bird sightings, cloud formations, or changes in swells and currents.[20]

So then, how accurate was latitude navigation? Late twentieth century experiments made aboard the *NRP Sagres* (a Portuguese Navy tall ship used as

---

[18] Which is far harder do determine astronomically, see Were San Martín's Longitude Measurements Accurate?.

[19] Using spherical trigonometry. Since pilots were not well versed in such calculations, pre-calculated tables were provided (Abel Fontoura da Costa, *A Marinharia dos Descobrimentos* […], on p. 48–143).

[20] None of these complementary navigation techniques were exclusive to European navigators. Polynesians, for instance, forwent instruments and relied exclusively on these methods for long-distance sailing in the Pacific (see Who Were the First to Cross the Pacific?)

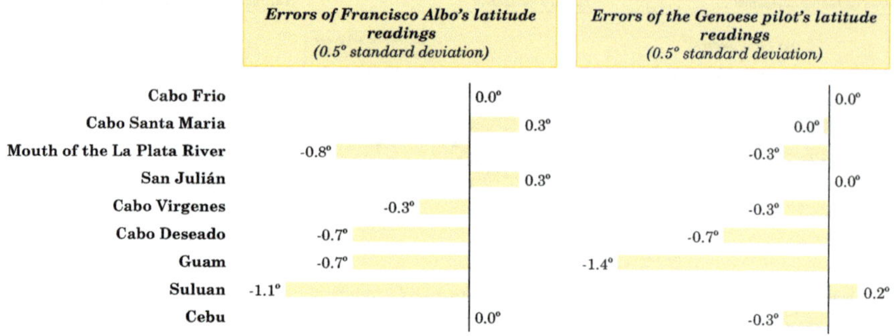

Fig. 3.5 Errors of latitude measurements made by Francisco Albo and the Genoese pilot for a sample set of locations[24]

a training vessel) with replicas of medieval quadrants and astrolabes suggest that these instruments measured angles with an error of about 0.4°.[21] The procedures used to convert these measurements into latitude added in the worst case another 0.5°.[22] The logbooks from Magellan's pilots show comparable levels of accuracy, with both Francisco Albo and the *Genoese pilot*[23] able to measure latitude with a standard deviation error of 0.5° (Fig. 3.5).

While this is incomparably more than the 5 m accuracy (0.00005°) of a GPS-enabled smartphone today,[25] latitude navigation was state-of-the art for the sixteenth century and would soon underpin several of the expedition's breakthroughs.

---

[21] José Malhão Pereira, "Experiências com instrumentos de navegação da época dos Descobrimentos", *Mare*
   Liberum, 7, p.165–192, on p.176.

[22] Abel Fontoura da Costa, *A Marinharia dos Descobrimentos* […], on p. 85.

[23] Likely León Pancaldo (José Manuel Garcia, "Documentos existentes em Portugal sobre Fernão de Magalhães e as suas viagens", *Abriu*, 8, 2019, p.15–33, on p. 20).

[24] Set of latitude readings from H.E.J. Stanley, *The first voyage round the world, by Magellan. Translated from the accounts of Pigafetta, and other contemporary writers* (London: Hakluyt Society, 1874), on p. 1–20, 211–236.

[25] Frank van Diggelen and Per Enge, "The World's first GPS MOOC and Worldwide Laboratory using Smartphones", *Proceedings of the 28th International Technical Meeting of the Satellite Division of The Institute of Navigation (ION GNSS+ 2015)*, 2015, p.361–369, on p. 361.

4

Sweet Rather Than Salty

From the Tordesillas Meridian (December 30th, 1519) to the La Plata River (February 3rd, 1520)

On January 10th, after about 10 days sailing southwest, the fleet sighted Cape Santa Maria, known today as Punta del Este in Uruguay. This was not quite yet unchartered territory for Europeans, as both Portuguese and Spanish expeditions had been there before.[1] Nautical charts placed the demarcation line here (shifted about 12° degrees west, due to magnetic declination[2]) so, with the confidence that he was finally in the Spanish hemisphere, Magellan started to thoroughly look for a passage (Fig. 4.1).

From Punta del Este, no further land is visible to the south, raising hopes the South American continent ended here. As they sailed west following the coastline, Magellan or one of his pilots is reported to have shouted *"Montem video!"* ("I see a mountain" in Latin) thereby christening Montevideo, Uruguay's capital.[3]

They continued to follow the coastline after Montevideo but noticed the water depth rapidly diminishing. Magellan ordered the *Santiago* and the *Victoria*—the two ships with the shallowest draught—to lead, taking frequent depth soundings. After three days and about 100 kms, little doubt remained that they were travelling up a river, as the water was now fresh and the depth had shrunken to less than 6 m. To be sure, Magellan ordered the *Santiago* to

---

[1] See Why Did Magellan Believe the Atlantic and the Pacific Were Connected?

[2] See How Did Magellan Convince the Spanish King?

[3] Visconde de Lagôa, *Fernão de Magalhães: A sua vida e a sua viagem* (Lisboa, Seara Nova, 1938), in Volume II, p.220, citing Pigafetta; José Toribio Medina, *El descubrimiento del Océano Pacífico: Vasco Núñez de Balboa, Fernando de Magallanes y sus compañeros* (Santiago de Chile: Imprenta Universitaria, 1920), on p.CCV, citing one of the pilots.

R. Gaspar, *The Revolution of Magellan*, https://doi.org/10.1007/978-3-032-10797-8_4

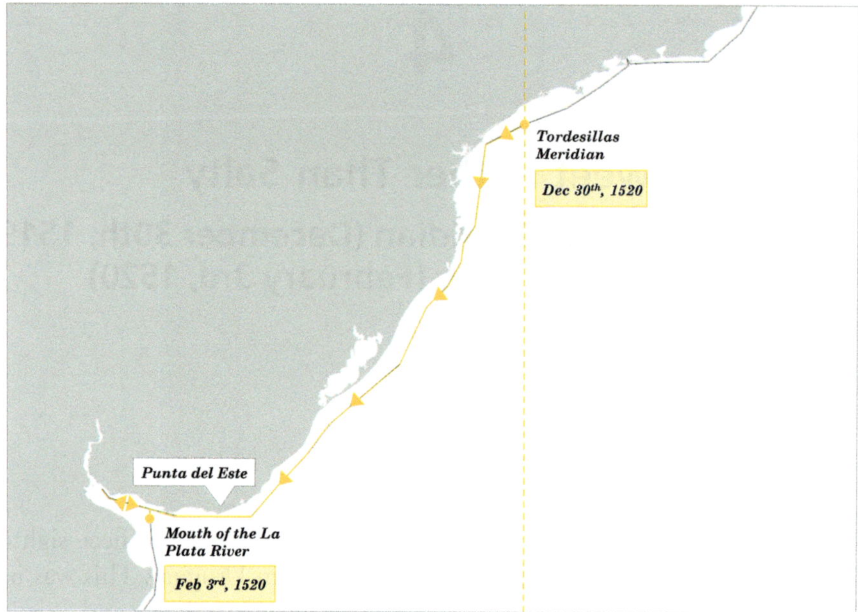

Fig. 4.1 Fleet route[4]

continue travelling upriver, while the rest of the fleet explored the south of what was apparently just an estuary.[5]

The southern exploration soon proved fruitless. While waiting for the *Santiago* to return, the crew turned their attention to the many natives visible along the shoreline, likely Querandi. Magellan, with what happened to Solis sharply imprinted in his mind,[6] had forbidden anyone to go ashore. One day though, a native paddled towards the ships in a canoe. Taller than Europeans and with a booming voice shouting angrily, he put the crew in guard but left before attempting to board any of the ships. Magellan, wary yet curious, sent a crew of 100 men—ten times more than the crew Solis had taken ashore— after him, but they returned emptyhanded.[7] A few days later, another one of the giant Querandis approached the fleet, and this time climbed aboard the *Trinidad*. Hoping to gain intelligence, Magellan first offered him a jacket of red cloth and other presents, and then showed him a silver platter. With hand

---

[4] Fleet route adapted from Tomás Mazón Serrano ("Mapas", Ruta Elcano, June 19, 2024, https://rut aelcano.com/mapas/); map adapted from Natural Earth (1:10 m large scale land data, version 5.1.1, https://www.naturalearthdata.com/downloads/).

[5] José Toribio Medina, *El descubrimiento del Océano Pacífico* […], on p.CCVI, CCVII.

[6] Solis was reportedly eaten by natives, see How Did Magellan Convince the Spanish King?

[7] José Toribio Medina, *El descubrimiento del Océano Pacífico* […], on p.CCVI, CCVIII.

gestures, the native seemingly indicated there was indeed silver in his land. He vowed to return, but never did.[8]

What did indeed return was the *Santiago*, bearing the dreaded confirmation they were indeed in an unusually large mouth of a river. In the fifteen days they had been gone, the *Santiago* had travelled more than 500 km up and down two arms of the river, confirming there was no passage.[9] It was a tough blow for Magellan's credibility, which in large part hinged on his knowledge of where to find a western passage.

4.1   Why Did Magellan Believe the Atlantic and the Pacific Were Connected?

The short answer

Magellan's belief in the existence of a passage through the Americas may have been based on faith, rather than fact. Despite confidently asserting to Charles V and his advisors the fleet would travel through "a certain strait he knew", his charts showed nothing beyond the La Plata River. His actions—ordering the ships to explore any nook and cranny of the South American coastline—were also those of someone not quite sure where a strait could be found, or if one existed at all.

Ptolemy wrote *Geography* in the second century, depicting the Indian and Pacific oceans as landlocked,[10] a notion that lingered in Europe for centuries. In 1488, the Portuguese explorer Bartolomeu Dias proved him wrong by rounding the Cape of Good Hope, showing that at least the Atlantic and Indian oceans were connected. Still, Ptolemy continued to be an influential figure, often used to fill in gaps about the yet unexplored parts of the globe. By 1519, when Magellan set sail, the Portuguese cartographer Lopo Homem authored a world map that extended the Americas to the conjectured Antarctica, landlocking the Pacific Ocean (Fig. 4.2). It was a patchwork of the new and the old, combining the recent discovery of the Americas with Ptolemy's nearly 1,400-year-old conjectures.[11]

---

[8] Jean Denucé, *Magellan: la question des Moluques et la première circumnavigation du globe* (Brussels: Hayez, 1911), on p.265.

[9] Tim Joyner, *Magellan* (Maine: International Marine, 1992), on p.132.

[10] Ptolemy, *Ptolemy's Geography: An Annotated Translation of the Theoretical Chapters,* edited and translated by Len Berggren and Alexander Jones (Princeton: Princeton University Press, 2000), on p.125–141.

[11] Joaquim Gaspar and Šima Krtalić, *The Cartography of Magellan* (Lisbon: Tradisom, 2023), on p.127–129.

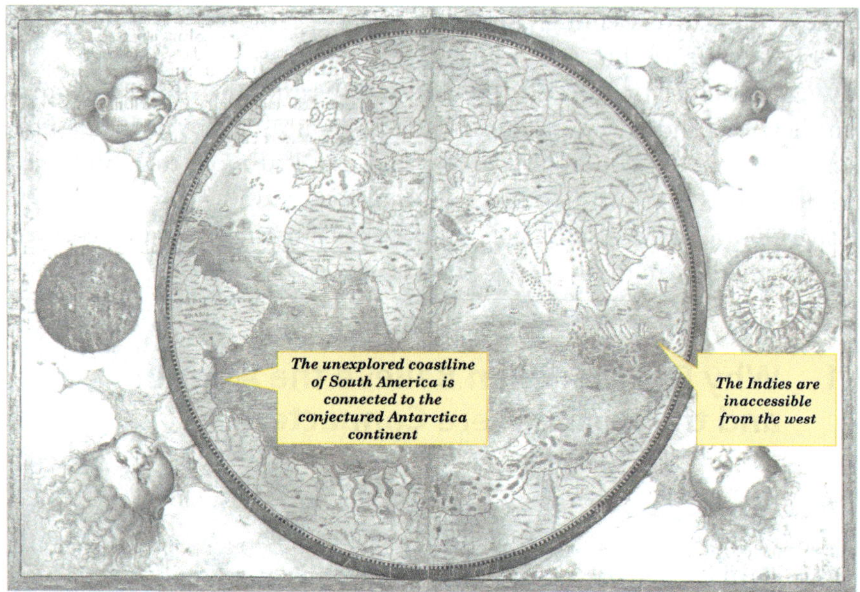

Fig. 4.2 Lopo Homem's 1519 world map suggested it was not possible to reach the Indies by travelling west[12]

Unlike Lopo Homem, Magellan excluded speculations from the chart he took to the Spanish king, which showed only the fragments of the American coastline that had been already explored. What could be hiding beyond that sliver of explored coastline?

In 1513 Balboa had explored the narrow Isthmus of Panama and seen a vast expanse of water beyond it, but had no way of knowing he was looking at the Pacific Ocean, or that there was a passage to it somewhere. Conversely, Magellan's promise to King Charles V was very clear, stating they would travel *"by a certain strait he knew"*,[13] located west of the demarcation line. Yet, his chart showed nothing south of the La Plata River (Fig. 4.3). While the need for secrecy may have played a role—after all, his value for the Spanish Crown would be greatly diminished if they knew where the passage was—it is altogether unclear if Magellan's knew where the passage was, or if one existed at all.

Magellan's main source of intelligence about this region may have been João de Lisboa, a Portuguese pilot he could have met en-route to his

---

[12] Lopo Homem, *Map of the Atlantic and Indian Oceans* in *Atlas Miller* (Paris: Bibliothèque Nationale de France, GED-26179 RES, 1519).

[13] Bartolomé de Las Casas, *Historia de Las Indias*, edited by Marqués de La Fuensanta Del Valle and D. José Sancho Rayon (Madrid: Imprenta de Miguel Ginesta, 1876), in Tomo IV, p. 296.

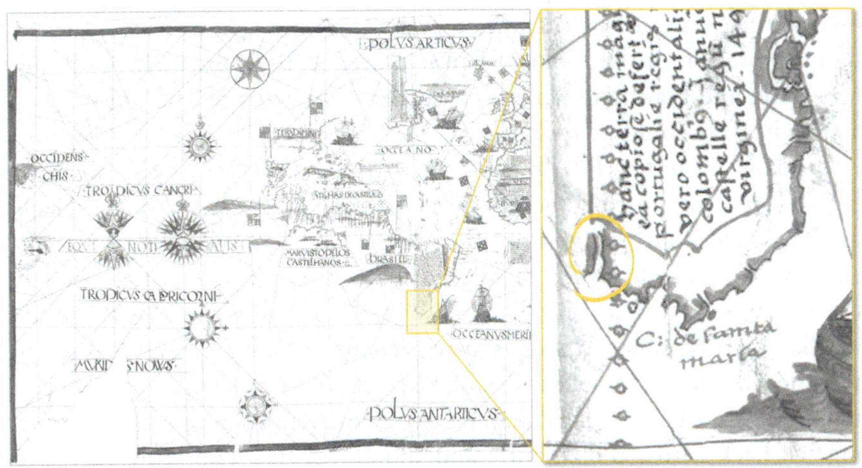

Fig. 4.3 The Kunstmann IV chart Magellan gave to the Spanish king, highlighting the mouth of the La Plata River as the last plotted location in the South American coastline[14]

campaign in northern Africa (João de Lisboa was the pilot-major of the ships) or later while living in Porto (where Magellan sought information from numerous pilots and merchants). Around 1512, an expedition piloted by João de Lisboa had reached the mouth of the La Plata River (later, in 1516, the ill-fated Solis would also explore it) and could have continued south as far as the Gulf of San Matias (about 7° south of La Plata).[15] Magellan's chart seems to incorporate the findings of João de Lisboa, as it includes a reasonably accurate description of the northern shore of the La Plata River, from Cape Santa Maria to the mouth of Uruguay River. However, it does not show the southern shore. This suggests either João de Lisboa did not explore this area thoroughly, or that there was a conscious decision not to incorporate all the surveyed areas in the chart. For instance, the Portuguese cartographers may have withheld information about the coastline west of the demarcation line as not to admit breaching the Treaty of Tordesillas and to avoid the risk of passing on intelligence to Spain.

Could João de Lisboa have provided Magellan with confidential and unpublished information? Even if he did, there is no evidence the Portuguese pilot ever went past the Gulf of San Matias, and certainly not all the way

---

[14] Attributed to Jorge Reinel and Pedro Reinel, *Kunstmann IV Nautical Planisphere* (facsimile by Otto Progel) (Paris: Bibliothèque Nationale de France, CPL GE AA-564, c1519).

[15] José Toribio Medina, *El descubrimiento del Océano Pacífico: Vasco Núñez de Balboa, Fernando de Magallanes y sus compañeros* (Santiago de Chile: Imprenta Universitaria, 1920, on p.LXV, LXVI.

down to the Strait of Magellan, more than 1,200 kms away. Magellan's thorough exploration of the La Plata River, as well as the subsequent probing of every nook and cranny down the coastline, further suggest he did not know where a passage could be found.

Magellan did have a Plan B, though. Bartolomé de las Casas, a missionary who was at court during the hearings with the king, later asked Magellan what he would do if he could not find a passage. He answered that *"he would go by the route the Portuguese followed"*.[16] The alternative is not as farfetched as it might seem at first. Magellan knew the Moluccas were close to the anti-meridian of Tordesillas, so the distance there from La Plata River was roughly the same whether he travelled west or east (see Fig. 1.3, Chap. 1). Spain could then establish a base in La Plata and use it to resupply the ships before the long sail back across the Atlantic and Indian oceans. The route was clearly less practical and more dangerous than the Portuguese one—which benefited from multiple resupply ports in Africa, India, and Malacca—but not altogether impossible.

Plan B had other issues, though. For one, it was not what he had promised the Spanish king. Two, it broke the rules the monarch gave him,[17] and Portugal would certainly complain the route broke the Treaty of Tordesillas. Three, it was always possible that a passage to the Pacific was hiding behind the next turn of the shoreline. Over the next months, Magellan would ignore Plan B and inexorably drive the fleet down the South American coastline. To the growing pleas of his officers to turn east and escape the cold, Magellan callously responded they would go as far as 75° (well within the Antarctic circle) to find a passage.[18]

---

[16] Bartolomé de Las Casas, *Historia de Las Indias*, [...], in Tomo IV, p. 296.

[17] "Instrucción dada por el Emperador a Fernando de Magallanes y Ruy Falero [...]" in Cristóbal Bernal, *Crónicas de la Primera Vuelta al Mundo, según sus Protagonistas*, 2016, on p.142–172.

[18] João de Barros, *Década terceira da Ásia* (Lisboa: Jorge Rodrigues, 1628), on f.142.

5

Into the Unknown

From the La Plata River (February 3rd, 1520) to Port San Julián (March 31st, 1520)

By February 3rd, the fleet had finished exploring the estuary of the La Plata River and was on its way back to the Atlantic Ocean, sailing in unchartered open waters for the first time (Fig. 5.1). It was hardly something to be happy about though, since searching for an unknown passage meant the ships would need to hug the coastline, a dangerous affair even on familiar waters. Without knowledge of the local winds, currents and shoals, the risk was even higher.

The fleet ran into problems soon enough. On February 10th, as the shoreline curved promisingly to the west, the ships were hit by a violent storm that dragged them back north towards the sand banks near Cape Corrientes. They dropped all anchors to avoid running aground, but the *Victoria* was not able to escape without dragging its keel a few times.[1] They were pushing their luck, so Magellan ordered the ships to sail away from the coast, dipping closer to shore only here and there to check for passages and places to resupply.

By February 28th they had rounded the cape of Punta Delgada and were met by *pamperos*, strong and cold winds which blow from the Andes, signalling the rapid approach of winter. With the ships battered and supplies dwindling, it was now clear they could not continue looking for a passage during the winter. The crew pleaded with Magellan to return to Rio de Janeiro but the captain general refused and kept the fleet moving south.[2]

Close calls became more frequent. After being separated by foul weather, the ships regrouped at a small anchorage they called *Bahía de Los Patos*

---

[1] José Toribio Medina, *El descubrimiento del Océano Pacífico: Vasco Núñez de Balboa, Fernando de Magallanes y sus compañeros* (Santiago de Chile: Imprenta Universitaria, 1920), on p. CCIX.

[2] Tim Joyner, *Magellan* (Maine: International Marine, 1992), on p. 134.

R. Gaspar, *The Revolution of Magellan*, https://doi.org/10.1007/978-3-032-10797-8_5

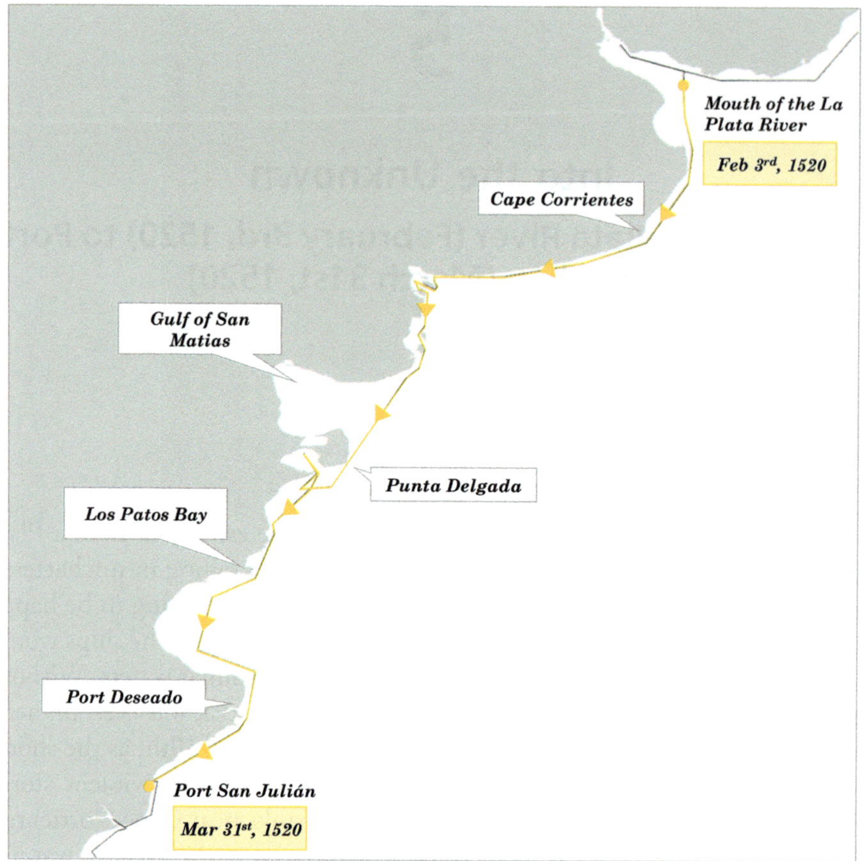

Fig. 5.1 Fleet route[3]

("Bay of the Ducks"), likely just south of modern-day Camarones Bay. These were of course not ducks freezing in the cold but rather penguins, which they had never seen before. Pigafetta preferred to call them "geese" and was also fascinated with the "sea-wolves" (seals) which could be seen everywhere.[4] Whatever they were they looked edible, and a crew of six men sent ashore to gather water and wood slaughtered a few. Before they could return, another tempest forced them to spend the night ashore. A search party sent the following morning feared the worst but spotted the six shivering men huddling among the seals, so as not to freeze to death. The joy of finding the

---

[3] Fleet route adapted from Tomás Mazón Serrano ("Mapas", Ruta Elcano, June 19, 2024, https://rutaelcano.com/mapas/); map adapted from Natural Earth (1:10 m large scale land data, version 5.1.1, https://www.naturalearthdata.com/downloads/).

[4] Antonio Pigafetta in H.E.J. Stanley, *The first voyage round the world, by Magellan. Translated from the accounts of Pigafetta, and other contemporary writers* (London: Hakluyt Society, 1874), on p. 49.

men still alive was short-lived, as a close sequence of gales prevented the fleet from moving on. For three days, the winds were so strong they blew away the forecastles of all five ships.[5]

The next anchorage, while offering more shelter, proved to be no less daunting, earning the name of *Bahía de Los Trabajos* ("Bay of Travail", now called Port Deseado). For six days, yet another squall prevented the crew sent ashore to return to the ships, who survived eating mussels.[6] The fleet continued to travel south and finally, on March 31st, found an anchorage with all four key elements required to sustain the fleet during the winter: shelter, water, food, and wood. They called it Port San Julián,[7] a name still standing today, and a place soon to become the stage for a dark winter.

5.1   How Advanced Were Magellan's Ships?

The short answer

Magellan's five ships were naus, which the Portuguese had started using a couple of decades before in the commercial routes to the Indies and Brazil. Compared to the earlier caravels, naus were less manoeuvrable but sturdier and able to carry more weight. Compared to modern ships though, they were small and fragile, difficult to operate, and exceedingly dangerous. Still, they were the most technically advanced machines of the 16th century, and as much state-of-the-art as spaceships today.

The fleet had made it to Port San Julián, but by the leanest of margins. One more squall, one more hit to the keel, or one more dragged anchor could have easily sent the ships crashing into the reefs or spiralling down to the bottom of the ocean. This begs the question: did Magellan have the right tools for the job?

His ships, called naus, had first been used for oceanic crossing by the Portuguese only a couple of decades before.[8] Rather than radical new creations, naus were a "best-of" compilation of naval technology. Faced with the challenges of establishing stable commercial routes to the Indies and

---

[5] José Toribio Medina, *El descubrimiento del Océano Pacífico* […], on p.CCX, citing Herrera.

[6] Idem, on p.CCXI, citing Herrera.

[7] Visconde de Lagôa, *Fernão de Magalhães: A sua vida e a sua viagem* (Lisboa, Seara Nova, 1938), in Volume II, p.222, citing the anonymous Genose Pilot and Hererra.

[8] One of the earlier references to their use is on Vasco da Gama's first voyage to India, in 1497 to 1499 (Francisco Contente Domingues, *Caravelas, Naus e Galeões – Séculos XV e XVI*, Lisboa, Caleidoscópio, 2017, on p. 40, 41).

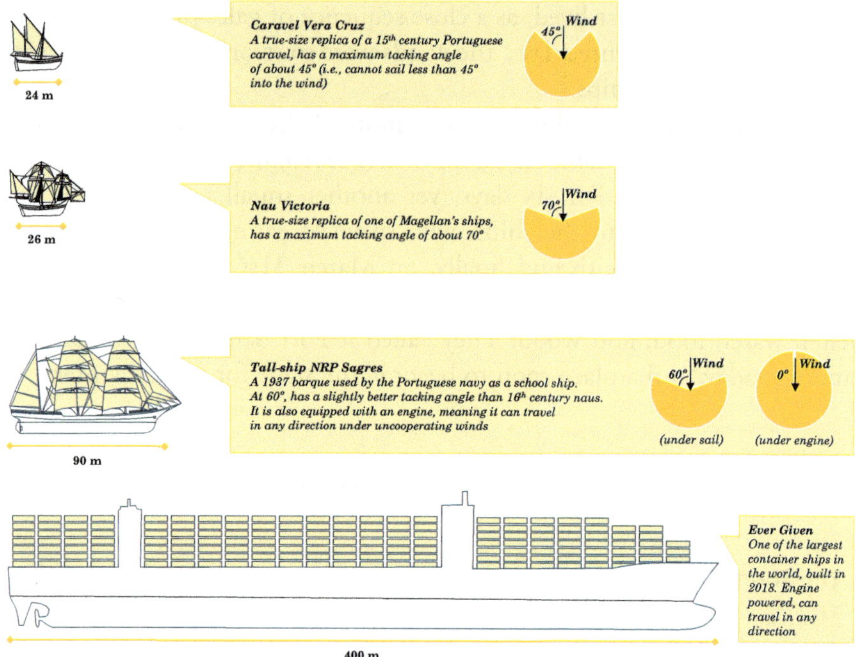

Fig. 5.2 Comparison of the dimensions and upwind sailing abilities of a caravel, nau, tall-ship, and a modern cargo ship[9]

Brazil, the Portuguese looked beyond the caravels used till then for ocean exploration. Whereas caravels were smaller and nimbler vessels, naus were larger and taller ships that traded agility for robustness and cargo space.

The increase in weight meant the triangular lateen sails used in caravels were no longer suitable. In alternative, naus used square sails which maximised hauling capacity at the expense of manoeuvrability and upwind sailing capabilities. The trade-off was a conscious one, as naus were meant to mainly sail downwind, leveraging trade winds on established commercial routes (Fig. 5.2).

To increase rigidity, rather than relying on the planks for structural integrity, the hull of naus was built around a lengthwise keel and width wise

---

[9] Adapted from: Simon Koppes, "Portuguese tallship NRP Sagres", CC BY 3.0, https://creativecomm ons.org/licenses/by/3.0/deed.en; "Vera Cruz", Aporvela, 2022, https://2023.aporvela.pt/2022/08/31/ veracruz/; Ignacio Fernández Vial y Francisco Monsalvete, "Replica Nao Victoria", *Museu Marítim de Barcelona*; Mary-Ann Russon, "The Cost of the Suez Canal blockage", BBC, 2021, https://www.bbc. com/news/business-56559073.

ribs.[10] The increased hauling capacity and rigidity also supported additional artillery, increasing its defensive and offensive capabilities.[11]

This combination of robustness, cargo capacity and heavy artillery would greatly aid sixteenth century Europeans in establishing and controlling global shipping routes.[12] However, Magellan was not sailing down a known route. The main trade-offs of naus—poor manoeuvrability and limited ability to travel upwind—would be significant shortcomings for his exploratory expedition, where most of the route and its prevailing winds were unknown. There was no other option though, as the more agile caravel did not have the robustness nor the space necessary to store provisions for such a long voyage. The caravel's reduced hauling capacity would also limit the trip's immediate economic benefits, as less spices could be brought back.

Magellan's five naus had been bought second-hand in Cadiz. By intent or by chance, no two ships were alike, providing the resulting fleet with a broad range of manoeuvrability, speed and hauling capacity. Such flexibility would mean however it would be harder to keep the ships together throughout long oceanic stretches. In any case, Magellan would have had a hard time finding five identical ships even if he wanted to, as there was little standardisation in sixteenth century shipbuilding. Extant documentation suggests there were no blueprints nor scale models to guide the construction process, only written rules. These mostly qualitative guidelines, which usually described only the construction of the ship up to the first deck, meant that no two naus were quite alike.[13]

As is still common today with freight ships, the key metric of a nau was its cargo capacity, and each of Magellan's five naus could carry between 75 and 120 tons.[14] A ton (*tonelada*) was the space required to store two casks, usually assumed to be around one cubic metre, or 1,000 litres.[15] Today, a ton is a measure of weight, not volume, so comparisons can be misleading. Caveats aside though, the scale of cargo ships has grown enormously: the *Ever*

---

[10] Francisco Contente Domingues, *Os navios do mar oceano: teoria e empiria na arquitectura naval portuguesa dos séculos XVI e XVII* (Lisboa: Centro de História da Universidade de Lisboa, 2004), on p. 231–233.

[11] Francisco Contente Domingues, *Caravelas, Naus e Galeões – Séculos XV e XVI* (Lisboa: Caleidoscópio, 2017), on p.41.

[12] See How Did Magellan's Expedition Change the World?

[13] Francisco Contente Domingues, *Caravelas, Naus e Galeões* [...], on p. 71–80.

[14] João Barata, "A Armada de Fernão de Magalhães", *Actas do II Colóquio Luso-Espanhol de História Ultramarina*, 1975, p. 109–134, on p. 111.

[15] Tim Joyner, *Magellan* (Maine: International Marine, 1992), on p. 248.

Given—the container ship that blocked the Suez Canal in 2021—can carry 200,000 tons.[16]

Apart from their tonnage, not much else is known about the physical dimensions of Magellan's naus. Based on sixteenth century Iberian guidelines for ship building, historians have estimated that the 75 to 120-ton naus were probably 15 to 30 m long,[17] not unlike a modern medium sized sailboat. Compared to modern cargo vessels though, naus were minuscule: the *Ever Given* is nearly 400 m long. Despite this size, it has a crew of only 25, not even half of what Magellan's naus required. Each of his sailors had less than three square metres of living space,[18] the equivalent of cramming 13 people into a 40 m^2 studio apartment.

Without present aids such as ancillary engines, electrical winches, and echo sounders, naus relied exclusively on the kindness of the winds, the skills of its officers, the brawn of their sailors, and a good dose of luck. Manoeuvres close to the coast were particularly taxing and dreaded, as they often required the intervention of the entire crew and the danger of grounding the ship was always looming.

Darkness made matters worse, so ships would wait for daytime before approaching land. To minimise the ship's movements, the crew would position the sails in such a way wind forces would mostly cancel each other. However, unless the wind was mild, the ship would still drift downwind, keeping the crew on its toes. This manoeuvre, called *heaving to*, was also used while waiting for other ships, launching a support vessel, or fetching a man who had fallen overboard.[19]

Entering or leaving a harbour was best done with following wind and incoming tide, so that the rising water would free the hull in case it got caught in a shoal.[20] A sailor would measure the water depth frequently, to mitigate the risk of grounding the ship. Depth soundings were taken using a simple yet remarkably effective device: a lead attached to the end of a long

---

[16] Mary-Ann Russon, "The Cost of the Suez Canal blockage" [...].

[17] Estimates vary widely and seem to be increasing in more recent analyses e.g.: Barata's 14 to 17 m ("A Armada de Fernão de Magalhães" [...], 1975, on p. 126); Joyner's 15 to 20 m (*Magellan* [...], 1992, on p. 248); Rossfelder's 15 to 28 m (André Rossfelder, *In pursuit of longitude: Magellan and the antemeridian*, La Jolla, Starboard Books, 2008, on p.159); the 26 m replica of the *Victoria* (Ignacio Fernández Vial y Francisco Monsalvete, "Replica Nao Victoria", *Museu Marítim de Barcelona*); or the 28 m estimate for the *Victoria* (Marcelino González Fernández, "La Nao Victoria", *Revista General de Marina*, August-Septmeber 2019, p.323–341, on p. 324).

[18] Adapted from Pablo Emilio Pérez-Mallaína Bueno, *Spain's Men of the Sea* (Baltimore: The Johns Hopkins University Press, 1998), on p. 131.

[19] José Malhão Pereira, "Goa, Daman and Diu, Safe Harbours for the Portuguese Trade", *Cities in Medieval India*, ed. Joseph Sharma and Pius Malenkadathil (New Delhi: Primus Books, 2014), on *Heaving To* section.

[20] Idem.

line. Some mariners would also place tallow on the lead to sample the seabed (sand bottoms were best for anchoring).[21]

Many harbours did not have docks large enough for naus, so they would need to anchor at a distance and use support vessels to go ashore. This was also the most sensible option at unfamiliar locations, as it would put some distance between the ship and potentially hostile locals. Still, dropping and raising the heavy anchors was a choreographed dance that could take up to 50 min. The ship needed to be moving backwards when the anchor hit the ground, requiring a close coordination between the sailor sounding the water depth, the one manning the helm, the crew operating the sails, and the group lowering the anchor. Botching the operation could mean losing the anchors (not an uncommon occurrence, as metal chains were not yet used and ropes were prone to snap under heavy loads) and seeing the ship dragged away by wind or current.[22]

Magellan's fleet included two different types of support vessels: longboats (or shallops), about seven metres in length, with two short masts and lateen sails; and skiffs (or launches), which were smaller row boats of about four meters. Skiffs were usually stowed in the main deck of the ships, while the larger longboats were towed to save space aboard.[23]

In open seas, the crew could breathe a little easier, at least until the master rallied them to adjust the position of the sails. On the earlier caravels, the triangular sail could be controlled with a single rope—called a sheet—attached to its lower corner. This technique is still used in most modern sailboats and is one of the reasons why even a reasonably sized vessel can be single-handed in lonely oceanic crossings. However, on a nau (and other modern large sailing vessels such as tall ships) the position of the larger square sails was controlled by ropes (called braces) attached to each end of a yard, plus sheets on the lower corners. Adjusting the position of a square sail required adjusting the ropes attached to each one of its corners, compared to adjusting the single rope of a triangular sail (Fig. 5.3).

The complexity of adjusting square sails meant changing the direction of a nau could sometimes take more than 30 min and require the full crew on deck, often even those who were sleeping after a long shift. This was one of the reasons why tacking—travelling upwind by moving in a zig-zag pattern—was avoided whenever possible. The other reason to avoid tacking was that

---

[21] "Lead line", *Oxford Reference*, retrieved January 2025, https://www.oxfordreference.com/display/10.1093/oi/authority.20111110201502756.

[22] José Malhão Pereira, "Goa, Daman and Diu, Safe Harbours for the Portuguese Trade" […], on *Entering harbour and anchoring* and *Leaving the harbour* sections.

[23] Tim Joyner, *Magellan* […], on p.248, 249.

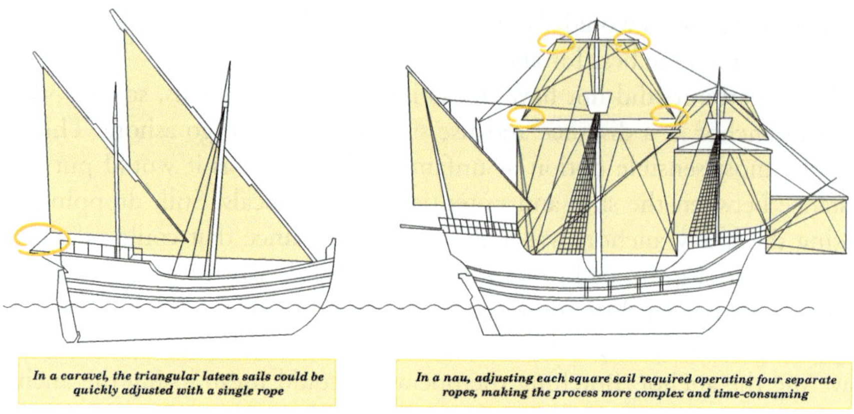

Fig. 5.3 Difference between adjusting sails in a caravel and in a nau[24]

the close-haul angle of a nau (the minimum angle at which a vessel can travel towards the wind) was quite large—around 70º—as its square sails could not rotate more than 30º before the yard would hit the standing rigging used to hold down the mast.[25]

As if operating a nau was not challenging enough, Magellan also had to contend with the difficulties of keeping a fleet together in open seas. Before they left Seville, the *Casa* had devised a set of rules to keep the fleet together. During daytime ships would follow the flagship *Trinidad,* which would cruise at a speed all ships in the fleet could match. At dusk, each ship would in turn steer close to the *Trinidad* and receive the orders for the night. To deal with unexpected situations, a set of light codes—made with wood torches inside lanterns—signalled orders to sail slower, tack, shorten sail, and be on the watch out for shoals. To ensure orders were understood, each ship was required to repeat the same light signal back to the *Trinidad.*[26]

The cost of acquiring the five naus second-hand was about 1,370,000 maravedies, and almost as much was spent extra in overhauling every hull, sail, and rigging (Fig. 5.4). Magellan and his officers seemed happy with the full overhaul, as no complaints were reported.[27] Later, while exploring

---

[24] Adapted from: "Vera Cruz", Aporvela, 2022, https://2023.aporvela.pt/2022/08/31/veracruz/; Ignacio Fernández Vial y Francisco Monsalvete, "Replica Nao Victoria", *Museu Marítim de Barcelona.*

[25] José Malhão Pereira, "Goa, Daman and Diu, Safe Harbours for the Portuguese Trade" […], on *The ships and its manoeuvre at sea* section.

[26] Charles McKew Parr, *So noble a captain: The life and times of Ferdinand Magellan* (New York: Crowell, 1953), on p. 61.

[27] Ignacio Fernández Vial, "Las Naves", *El viaje más largo: La primera vuelta ao mundo* (Madrid: Sociedad Mercantil Estatal de Acción Cultural, 2019), on p. 77.

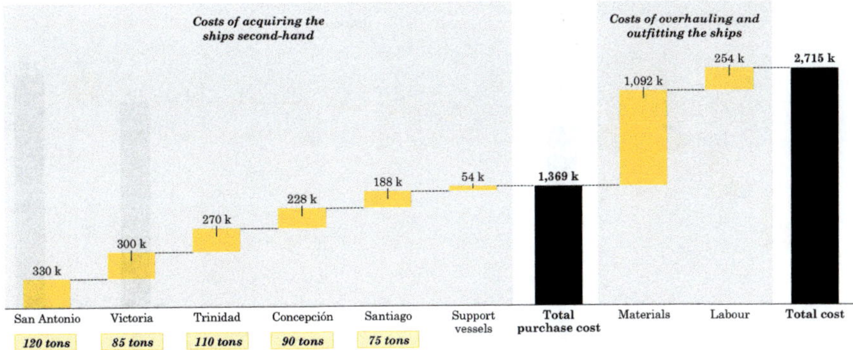

Fig. 5.4 Overview of the costs of acquiring and overhauling the fleet's ships, in thousands of maravedies[29]

the South American coastline, one of them thanked God for the "good and well-rigged ships".[28]

A sailor in the fleet earned 14,400 maravedies per year[30] so, on average, each ship cost 30 yearly salaries. This is certainly inexpensive by today's standards, considering that each team participating in the America's Cup reportedly spends north of €90 million,[31] and the United States' endeavour to reach the Moon required €260 billion.[32]

So, by modern standards, the fleet's ships were cheap, small, fragile, difficult to operate, and exceedingly dangerous. The comparison is of course an unfair one, as any 500-year-old advancement will pale in comparison to what mankind has achieved by the twenty-first century. By sixteenth century standards though, naus were the most technically advanced machines of their time, on water or on land.[33]

[28] Martín Fernández de Navarrete, *Coleccion de los viajes y descubrimientos que hicieron por mar los españoles desde fines del siglo XV* (Madrid: Madrid: Biblioteca de Autores Españoles, Ediciones Atlas, 1964), in Tomo LXXVI, p. 440.

[29] Adapted from: Lourdes Díaz-Trechuelo, "La organizacíon del viaje Magallânico", *Actas do II Colóquio Luso-Espanhol de História Ultramarina, 1975*, p.265–314, on p.274; Ignacio Fernández Vial, "Las Naves", *El viaje más largo* [...], on p. 77.

[30] See What Drove Men to a Life at Sea?

[31] Casey Newton, "Billionaire death race: inside America's Cup and the world's most dangerous sailboat", The Verge, 2013, https://www.theverge.com/2013/9/3/4672686/billionaire-death-race.

[32] "How much did the Apollo Program cost?", *The Planetary Society*, 2020, https://www.planetary.org/space-policy/cost-of-apollo.

[33] Henrique Leitão, "All Aboard!: Science and Ship Culture in Sixteenth-Century Oceanic Voyages", *Early Science and Medicine*, 21, 2016, p.113–132, on p. 122.

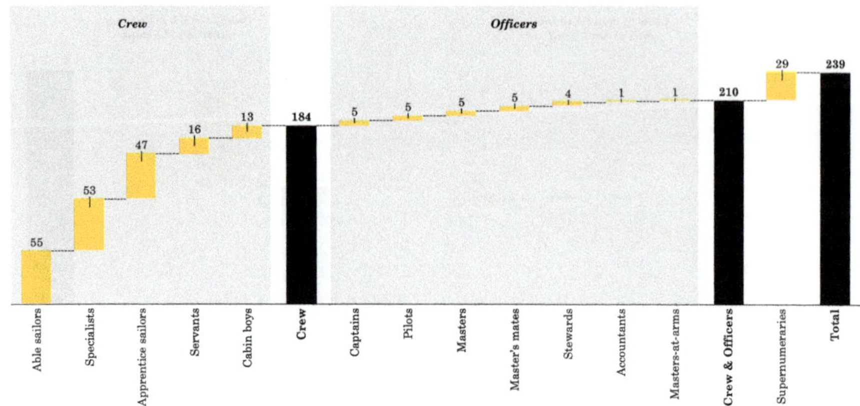

Fig. 5.5 Split of the fleet's mariners by rank,[35] excluding an estimated ~ 30 additional mariners which were not part of the *Casa's* original rosters[36]

5.2   What Was the Life of a 16th Century Sailor Like?

The short answer

A ship was a cramped, smelly, and dangerous place. Such aggravations applied to everyone, but not quite to the same degree. For the sailors and other members of the ordinary crew who filled up most of the positions, life aboard was harder than for the officers. They usually started as young cabin boys, and slowly and perilously made their way up. The smartest and shrewdest could aspire to become lower ranking officers, although the highest position of captain was outside the reach of a commoner. The small comforts possible aboard a ship—from sheltered sleeping quarters to more secluded latrines—were also off-limits.

For the sailors and other members of the regular crew who made up nearly 80% of the rosters (Fig. 5.5), life aboard was quite different from that of officers. Still, the gap was not as wide as it was on land, as the close quarters led to more relaxed social norms[34] and the dangers of the sea were blind to rank. Everyone was literally on the same boat.

For many, a mariner's career started at the very bottom of the ladder, as a cabin boy. The youngest ones were about seven years old, and the oldest

---

[34] Henrique Leitão, "All Aboard!: Science and Ship Culture in Sixteenth-Century Oceanic Voyages", *Early Science and Medicine*, 21, 2016, p. 113–132.

[35] Adapted from Tim Joyner's ship rosters: *Magellan* (Maine: International Marine, 1992), on p. 252–263.

[36] See What Drove Men to a Life at Sea?

eighteen. Apart from a select few—for instance, the protégé of a senior officer being groomed for higher positions—cabin boys had a rough introduction to life at sea. They received orders from any member of the crew and thus spent most of their days occupied with thankless tasks nobody wanted to do, from cleaning the ship to distributing provisions to picking up after meals. Cabin boys learned the art of sailing both by watching older crew members and by keeping watch during the night shifts. They were also responsible for turning over the sand clocks—the only time-keeping device aboard—every thirty minutes. Each time they recited a short religious hymn to let the shift officer know they had not forgotten nor fallen asleep.[37] Cabin boys were also usually the ones reciting the daily prayers, only grudgingly accompanied by the sailors as many were not fervorous followers of religious precepts.[38]

Cabin boys were customarily considered too young to take a beating, so that burden fell instead onto the backs of apprentice sailors. Most were between 17 and 20 years old, and often found themselves suffering at the hands of frustrated, angry, and drunken sailors. Beatings and non-sanctioned physical punishments were supposedly forbidden, but the rule was seldomly strictly enforced. Apprentices also saw their young vigour exploited for the most dangerous tasks aboard—climbing up the yards to furl the sails or keep watch on the crow's nest, regardless of the weather—and to load and unload heavy cargo. Some also rendered personal services to older crew members in exchange of small gifts or protection. These services ranged from preparing meals and making beds, to cleaning toenails and shaving legs.[39]

At the age of 20 or so, the competent apprentice raised to the rank of able sailor, receiving from the pilot or the master a signed document certifying his experience. While tasks were now slightly less physically demanding, the harsh seas still took their toll, so it was rare to see sailors more than 40 years old. Critical duties were entrusted to the more experienced sailors, such as manning the tiller (the wheel, which made steering a large ship easier, had not yet been invented) and handling the rigging during complex manoeuvres. In an all-too-common vicious circle, the apprentices who had endured years of physical and verbal abuse from older sailors were now the ones dispensing it.[40]

One possible step up in the ship's hierarchy was to become a carpenter (who maintained all wooden structures aboard, including the hull, mast, and

---

[37] Pablo Emilio Pérez-Mallaína Bueno, *Spain's Men of the Sea* (Baltimore: The Johns Hopkins University Press, 1998), on p. 75–77.

[38] Idem, on p. 239.

[39] Idem, on p. 77, 78.

[40] Idem, on p. 78, 79.

pulleys), a caulker (used oakum and tar to keep the hull and deck watertight, and kept the bilge pumps in working order), a gunner (made powder, operated and maintained the cannons), or a barber-surgeon (combining trimming and medical duties was very common in the Middle Ages and beyond). There was usually only one craftsman of each kind aboard, so the role paid better and was more prestigious than that of a sailor. However, it was not easy for a sailor to become a specialist, since he would need to find the time to learn the trade and, in the case of a gunner, also have the funds to pay for the *Casa's* required course and exam.[41]

An alternative way to raise through the ranks was to become an officer. While a smart and hardworking sailor could aspire to become a pilot (if he survived long enough to accumulate experience and pass the *Casa's* exam) or perhaps even a master (easier if he belonged to a lineage of respected mariners), the top position of captain was unattainable to the masses. Gonzalo Goméz de Espinosa, initially a master-at-arms in the *Trinidad* and one of the expedition's survivors, would raise to the position of captain in the later stages of the voyage. When he finally returned to Spain, after eight years of privations, the king refused to pay him the salary of captain, arguing that a mere master-at-arms, someone who deals with brute force and manual labour, does not have the eminence and education of a commander.[42]

The stigma of manual labour—and of the resulting tanned countenance and callous hands—would continue long after Magellan's expedition. In the early seventeenth century, spurred on by the need to increase the number of Spanish mariners, King Philipp III issued a decree reassuring low-ranking nobles that serving on a ship would not strip them of their title.[43] However, the king's decree did not seem to do much to increase the nobility of seafaring, as most low-ranking nobles still preferred to take their chances on a military career, starting as simple soldiers in hopes of rising through the ranks.[44]

The lower-tiered gentlemen willing to make a life at sea usually started as supernumeraries (as Magellan had done[45]). Most of them had military training and were thus called to action in armed excursions. When aboard,

---

[41] Idem, on p. 79–81.

[42] "Autos de Gonzalo Gómez de Espinosa: abono de salario" (Seville: Archivo General de Indias, ES.41091.AGI//PATRONATO,35,R.1, 1528), https://pares.mcu.es/ParesBusquedas20/catalogo/descri ption/122241.

[43] "Expediente sobre imprimir las ordenanzas y privilégios concedidos a la gente de mar de la armada de la Carrera de las Indias" (Seville: Archivo General de Indias, ES.41091.AGI// CONTRATACION,72, 1606).

[44] Pablo Emilio Pérez-Mallaína Bueno, *Spain's Men of the Sea* [...], on p. 36.

[45] See Why Was Magellan, a Portuguese Citizen, Serving Spain?

they took their meals and shared living quarters with the crew, but their noble standing provided a fast-track to ascend to officer positions, provided they were smart enough to learn the trade.

The lowest officer position was that of the steward, who was responsible for guarding and dispensing rations. The role seems perhaps trifling, but being on the good side of the steward could spell the difference between eating well or not eating at all, especially when supplies dwindled. Next came the master's mate, a sergeant of sorts, responsible for enforcing the orders given by the master, the overseer of day-to-day operations. During manoeuvres, the master stood at the poop and his mate at the prow, checking if sailors followed orders in a coordinated fashion. Due to the size and importance of Magellan's fleet, some of the tasks which habitually also befell on the master and his mate were taken by others. For instance, there were an inspector general and a fleet accountant (nominated by the Crown to protect its monetary interests), and a master-of-arms (responsible for maintaining discipline and dispensing punishment).[46]

The pilot's ranking was on par with that of the master, but it came with different responsibilities. While the latter was responsible for operations, the former took charge of navigation. Pilots were often the oldest aboard and required a formal education (to read and write, operate nautical instruments, and perform calculations), experience at sea (to interpret the conditions of the sea and the winds, and recognise the coastline), and a certification issued by the *Casa*. This had not always been the case, though. In the early days of Iberian oceanic exploration, pilots were men of paltry education and social standing, but the difficulty and criticality of their role grew as ships ventured deeper into uncharted territory.[47] Magellan's fleet included five pilots and one of them was also an astronomer—Andrés de San Martín, responsible for longitude measurements.[48]

Finally there was the captain, the highest authority on board. Towards the end of the sixteenth century, when transoceanic crossings became commonplace, the position of captain on a merchant vessel was mostly honorific. Unless the ship was attacked—in which case he would coordinate the defence—the captain would leave the running of the ship to the master and pilot.[49] However, in riskier exploratory expeditions, the captain held the responsibility for all key decisions, and that was certainly the case for Magellan.

---

[46] Pablo Emilio Pérez-Mallaína Bueno, *Spain's Men of the Sea* [...], on p. 81, 82.

[47] Idem, on p.83–87.

[48] See Were San Martín's Longitude Measurements Accurate?

[49] Pablo Emilio Pérez-Mallaína Bueno, *Spain's Men of the Sea* [...], on p. 95, 96.

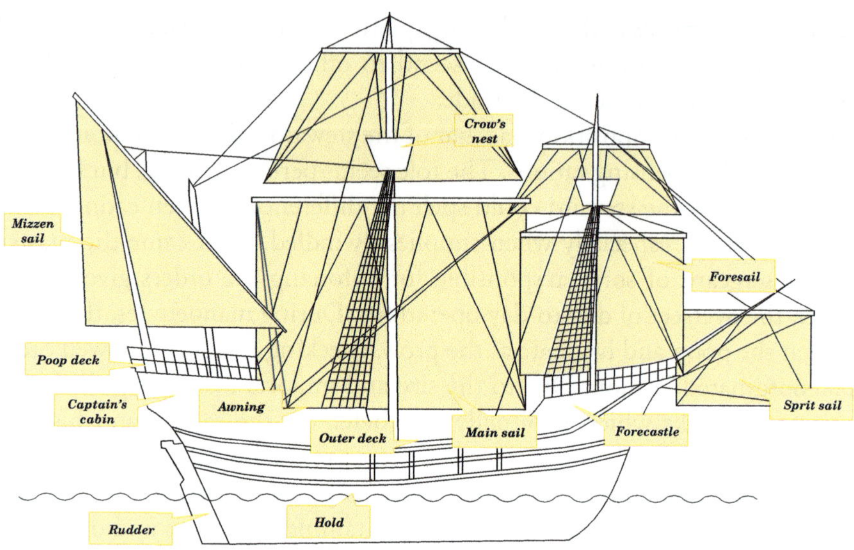

Fig. 5.6   Main areas in a nau[52]

Regardless of rank and social status, all those aboard were confined to the same cramped space for months on end. The foul smell of unwashed bodies and livestock made its way up even the most distinguished nostrils. Rats, cockroaches, lice, and bedbugs made no social distinction either. At the start of the journey, seasickness would also overwhelm rich and poor, experienced and novice alike.[50]

Social status still counted, though. While everyone had to endure the discomforts of sleeping on an overcrowded, swaying, and draughty ship, some had it better than others. The captain—and sometimes also the master and the pilot—had his own cabin, situated at the *toldilla* (poop deck), the most stable part of the ship (Fig. 5.6). Lower ranking officers slept beneath the *tolda* (awning) or the less stable but still sheltered *castillo* (forecastle). Everyone else looked had to find a nook to lay down their sleeping pad. In mild weather, many mariners preferred to sleep in the outer deck rather than endure the pestilence of the inner deck.[51]

Bowel movements were also not quite the same for everyone. Officers normally had more secluded latrines at the poop, while everyone else made their business at the prow, on top of a wooden grating protruding over

[50] Pablo Emilio Pérez-Mallaína Bueno, *Spain's Men of the Sea* [...], on p. 135.

[51] Idem, on p. 137.

[52] Drawing adapted from Ignacio Fernández Vial y Francisco Monsalvete, "Replica Nao Victoria", *Museu Marítim de Barcelona*.

Fig. 5.7 Left: Portrait of Charles V, showcasing the typical outfit of a nobleman[55]; Right: Drawing of a captain, showing the looser garments used aboard[56]

the water.[53] Today's precepts of privacy are stronger than they were in the sixteenth century, but the morning routine of exposing one's bottom to the ocean and to everyone else was certainly not the day's highlight.

Then there was the matter of clothing. The rigours of life at sea meant both officers and sailors wore garments that were looser, more comfortable, and less adorned than what was dictated by sixteenth century fashion. In lieu of breeches (balloon styled trousers cut above the knees and continuing down to the feet with fitted stockings), mariners used *zaragüelles* (loose fitting pants that fell from the waist to the knees). Rather than a doublet (snug-fitting short jacket used over the shirt), they wore long and thick sea capes which offered protection from the elements[54] (Fig. 5.7).

---

[53] Pablo Emilio Pérez-Mallaína Bueno, *Spain's Men of the Sea* [...], on p. 140.

[54] Pablo Emilio Pérez-Mallaína Bueno, *Spain's Men of the Sea* [...], on p. 145–152.

[55] Jakob Seisenegger, "Portrait of Emperor Charles V with Dog (1500–1558)" (Kunsthistorisches Museum, Inv.-Nr. GG A114, 1532).

[56] Adapted from Christoph Weiditz, "Trachtenbuch" (Germanisches Nationalmuseum, 1530–1540), on p. 86.

At a distance, the outfits of officers and sailors looked similar, but the illusion broke down at closer range. The first distinctive element was the hat: sailors wore a simple woollen red *bonete* that merely protected their heads from the cold, wealthier officers had more elaborated caps which offered both style and substance. Quality and finish varied considerably, from ill-fitted garments made from coarse linen to carefully tailored outfits made from velvet, taffeta, and satin. A peek inside the sea chests storing the personal belongings of mariners would also reveal stark differences in quantity. Most sailors had no more than a pair of shoes and one or two sets of clothing, but officers complemented their wardrobe with additional outfits to use in land.[57] For Magellan and his high command, the ability to swap their comfortable sailing garments for richer attire was an important part of their plan to impress foreign rulers.

---

[57] Pablo Emilio Pérez-Mallaína Bueno, *Spain's Men of the Sea* [...], on p. 145–152.

6

The Uprising
Port San Julián (March 31st, 1520 to April 3rd, 1520)

Even under the relative shelter of Port San Julián and with access to food, water and wood, the winter would be long and arduous. After the numerous close calls, the ships were battered and the crew demoralised. Magellan decided to reduce rations of biscuit and wine,[1] which made matters worse in two ways. One, the crew would need to endure the cold without the comforts of wine (which was not a bad thing entirely, as wine is a vasodilator that gives a false feeling of warmth). Two, it meant Magellan had no intentions of returning to Spain as many had pleaded, and was instead preparing to resume the search for a western passage as soon as the winter passed (Fig. 6.1).

More than six months had passed since the fleet left Spain, and Magellan still refused to share any details about his route to the Moluccas. His silence probably provoked a broad array of interpretations. Some still believed he was an experienced navigator with unique insights from his previous dealings with the Portuguese. Others had taken the erratic explorations after they left Rio de Janeiro as proof he was just a madman with no idea of where they were going. His worst detractors questioned not only his competence but also his honour, believing he was secretly working for Portugal. According to this rather convoluted theory, his intention was to sink the ships, maroon the survivors, and return to the good side of the Portuguese monarch.[2]

---

[1] Jean Denucé, *Magellan: la question des Moluques et la première circumnavigation du globe* (Brussels: Hayez, 1911), on p. 271.

[2] João de Barros, *Década terceira da Ásia de João de Barros* (Lisbon: Jorge Rodriguez, 1628), on Livro V, Capítulo IX, f. 142.

R. Gaspar, *The Revolution of Magellan*, https://doi.org/10.1007/978-3-032-10797-8_6

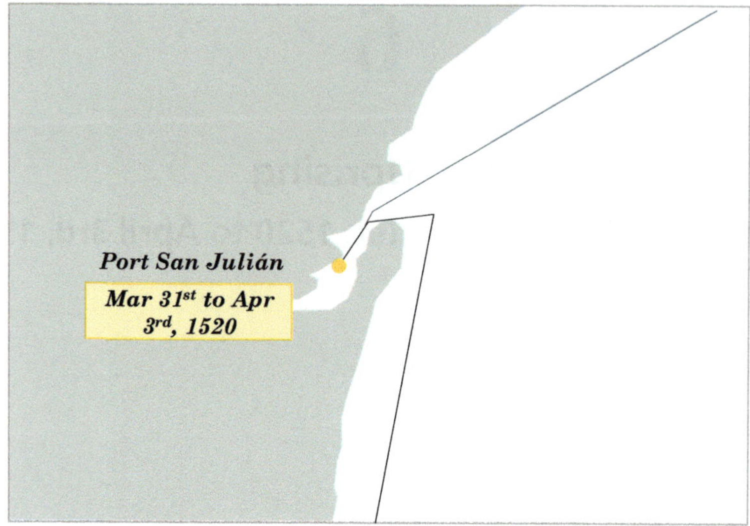

Fig. 6.1 Fleet route[3]

The truth was likely more straightforward, although it also did not paint Magellan in a particularly favourable light. They were now close to 50° latitude, 10° below the mouth of Rio de La Plata, and he had not yet found a western passage. If he returned to Spain empty-handed he would be left poor, disgraced or, worse still, in prison for defrauding the king. Magellan was bent on finding a passage further south or die trying. He was not alone though, and his decision affected the fate of hundreds of men.

The Spanish captains wanted to get to the bottom of Magellan's motives. Their faction was now led by Gaspar de Quesada, the captain of the *Concepción*, to whom Magellan had entrusted Cartagena after the second mutiny attempt at Rio de Janeiro. Quesada and Cartagena met with Mendoza—the captain of the *Victoria* – and other officers, to decide what to do.[4]

Among those officers was Juan Sebastían Elcano, serving as the master for the *Concepción*. Elcano was 33 years old and an experienced professional mariner. Born near San Sebastían, the Basque quickly rose through the ranks to become the master of a 200-ton ship (more than double the carrying capacity of the *Concepción*) chartered to the Spanish Crown. His reputation suffered after he was forced to sell the ship to pay off debts, but he

---

[3] Fleet route adapted from Tomás Mazón Serrano ("Mapas", Ruta Elcano, June 19, 2024, https://rutaelcano.com/mapas/); map adapted from Natural Earth (1:10 m large scale land data, version 5.1.1, https://www.naturalearthdata.com/downloads/).

[4] José Toribio Medina, *El descubrimiento del Océano Pacífico: Vasco Núñez de Balboa, Fernando de Magallanes y sus compañeros* (Santiago de Chile: Imprenta Universitaria, 1920), on p.CCXIII, CCXIV.

still managed to enlist in Magellan's expedition.[5] Elcano would play a vital role later in the voyage but, for now, he was just one more Spanish officer disgruntled with Magellan's leadership.

The Spaniards agreed to send a letter to Magellan, urging him to reveal the way to the Moluccas. If he did not comply, they would remove him by force.[6] To make their demands more compelling, the plotters planned to take over the *San António*, now captained by Álvaro de Mesquita, Magellan's kinsman. To avoid putting themselves again in the lion's den, they would skip Magellan's invitation for officers to have Easter dinner aboard the *Trinidad*.[7]

That night, following an awkward Easter dinner at the *Trinidad* with Mesquita as the only guest,[8] a longboat silently bridged the waters between the *Concepcíon* and the *San António*. It carried Quesada and Cartagena, plus 30 armed men. The night watch was either sleeping or was swiftly subdued, as no alarm sounded when the intruders climbed aboard the *San António*. They quickly made their way to the poop deck and arrested Mesquita, sound asleep in his cabin. Refusing to join the uprising, he was shackled and confined to his quarters. In the meantime, Juan de Elorriaga, the *San António's* master, woke up and confronted the mutineers, demanding they return to the *Concepcíon*. Rather than complying, Quesada stabbed him multiple times, leaving the master bleeding on the floor (he would eventually die from his wounds, several months later).[9] The first casualty of the uprising was not Magellan or any of his countrymen, but a Spaniard, sending a clear signal of what could happen to further dissidents. Still, the crew did not seem eager to join the rebellion, prompting a new round of arrests which included Diego Hernandéz—the master's mate—and Juan Rodriguéz de Mafra—the pilot, after he refused to replace Elorriaga as the master.

Alas, the takeover had been more hostile than they had hoped for. To win back the crew and signal they would no longer be subject to Magellan's ill treatment, they ordered Juan de Gopeguy—the steward—to lift the restrictions and let the hungry sailors eat as much as they wanted to. Knowing those rations would be needed later, whether they continued the voyage or returned to Spain, Gopeguy complained but Quesada forced him to comply under a barrage of threats.[10]

---

[5] Idem, on p.CCCLXV-CCCLXX.

[6] José Toribio Medina, *Colección de documentos inéditos para la historia de Chile: desde el viaje de Magallanes hasta la batalla de Maipo: 1518–1818* (Santiago de Chile: Imprenta Ercilla, 1888), in Tomo I, p.165, 301.

[7] José Toribio Medina, *El descubrimiento del Océano Pacífico* [...], on p.CCXIV, CCXV.

[8] Jean Denucé, *Magellan* [...], on p. 272.

[9] José Toribio Medina, *El descubrimiento del Océano Pacífico* [...], on p. CCXVI.

[10] Idem.

The Spanish officers now controlled the *San António*, the *Victoria*, and the *Concepcion*, much as they had on the first mutiny attempt, while crossing the Atlantic. And, once more, they hesitated to use the superior fire power to take over the flagship. Instead, they sat down to write a letter to Magellan.[11]

At the break of dawn, the *Trinidad* sent a skiff to the *San António* to hail a couple of sailors for the routinary trip ashore to collect water and wood, where they were warned by the crew that the *San António* was now in the hands of Quesada. After hearing the troubling news, Magellan immediately sent back the skiff to all ships to assess the extent of the uprising. The *San Antonio*, the *Concepción* and the *Victoria* declared they were loyal to King Charles V and nobody else. In the small *Santiago*, anchored nearby the *Trinidad*, nobody had yet realised there was a mutiny and a confused Serrano declared himself loyal to Magellan. Serrano, despite being Spanish, had shown little interest in the previous rebellion attempts.[12]

Later in the day, the *San António* longboat went to the *Trinidad* to deliver the letter the Spanish officers have been preparing. Perhaps reluctant of committing to writing outright mutiny, the Spaniards downplayed their actions. They had seized three ships merely to stop Magellan from continuing to mistreat and endanger the crew, and would return them if the captain general agreed to share the route and consult them in all important matters, as the king had ordered. If Magellan did so, they would gladly submit back to his leadership and kiss his hands and feet.[13]

It was an odd letter, even for a time when official communications were more grand sounding than they are today. It was also incoherent: if all they wanted was to be consulted as peers, why would they kiss Magellan's feet? And how did the stabbing of Elorriaga fit with their goal of protecting the crew? Magellan likely picked up on the lack of confidence of his opponents but was also not willing to put in writing anything that could later be construed as tyrannical. His next move was to invite the Spanish officers to the *Trinidad*, to discuss matters properly. He promised them fair treatment but, not surprisingly, they refused and insisted instead on meeting aboard the *San António*.[14]

Magellan also thought that risky, so the two factions were at a deadlock. As time passed, the likelihood of the three mutinous ships deserting or attacking the flagship increased, so Magellan needed to act fast. The crew of the *San*

---

[11] Idem, on p. CCXVII.

[12] Antonio Herrera, *Historia general de los hechos de los castellanos en las islas i Tierra Firme del Mar Oceano* (Madrid: En la emplenta Real, 1601), on Decada II, Libro IX, p. 298.

[13] José Toribio Medina, *El descubrimiento del Océano Pacífico* [...], on p. CCXVII.

[14] Jean Denucé, *Magellan* [...], on p. 273.

António longboat had spent the day back and forth delivering messages. They were most probably tired, cold, and hungry, so Magellan invited them aboard for a rest and a good meal. While they ate, under the cover of the dwindling daylight, Magellan's men pilfered the longboat and stored it out of sight.[15]

With the communications to the *San António* severed, Magellan discreetly sent two sailors and Espinosa—his master-at-arms—to the *Victoria*. Their official mission was to deliver a message to Captain Mendoza, but the daggers concealed under their sea capes suggested otherwise. A suspicious Mendoza was not willing to let Espinosa come near him, but after the master-at-arms questioned his nerve, the captain allowed him and just one of the sailors aboard. An unarmed Mendoza escorted the two men to his cabin, where Espinosa handed him a letter from Magellan. The letter summoned him to the *Trinidad,* to which Mendoza promptly replied he would not repeat that mistake. These would be his last words, as Espinosa and the sailor quickly reached for their hidden daggers and stabbed him in the throat and head, killing the captain.[16]

In the meantime, Magellan had loaded the confiscated longboat from the *San António* with fifteen armed men. Taking advantage of the incoming tide, they drifted silently towards the *Victoria* and, at a signal from Espinosa confirming the captain had been taken out, they stormed aboard and overpowered the deck watch. The crew, chiefly made of foreigners (see Fig. 2.2, Chap. 2), offered no resistance.[17]

As soon as the tide turned, the *Victoria,* now with Magellan's family banners on the mast, raised anchors and drifted towards the *Trinidad* and the *Santiago.* The three ships were positioned abreast at the entrance of the bay, trapping the *San António* and the *Concepción* inside (Fig. 6.2). The two rebellious ships still had superior firepower and could have perhaps used the cover of darkness and outgoing tide to force their way of the harbour, but it seems panic had already taken hold of the mutineers. Aboard the *San António,* Mesquita was released and asked to intercede in their favour, which he flatly refused.[18]

During the night, the *San António* started drifting with the current towards Magellan's ships. It seems Quesada, to expedite an attack or an escape as soon as there was enough light, had ordered two anchors to be lifted, hoping that the third one would hold the ship steady. It did not, and now the *San*

---

[15] Tim Joyner, *Magellan* […], on p. 140.

[16] José Toribio Medina, *El descubrimiento del Océano Pacífico* […], on p.CCXVIII.

[17] Tim Joyner, *Magellan* […], on p. 141.

[18] Idem, on p. 141.

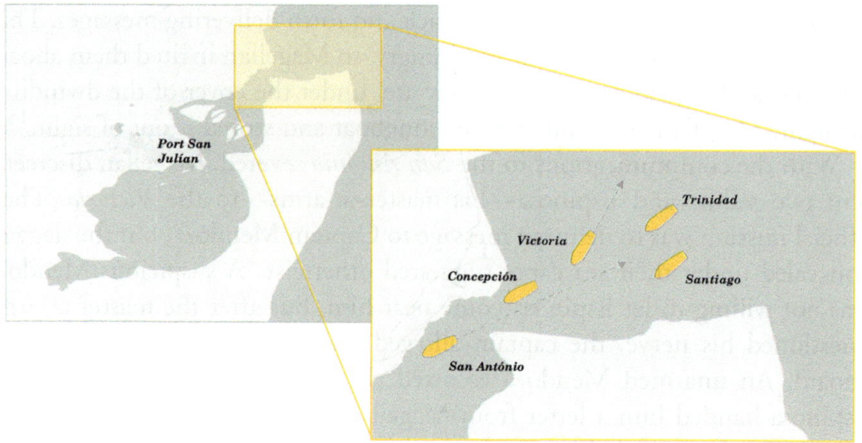

Fig. 6.2 Probable position of the ships at the start of the mutiny attempt, and subsequent movements of the *San António* and the *Victoria*[19]

António was drifting towards the entrance of the harbour, while the *Concepción* remained securely anchored.[20] Quesada jumped to the deck in full armour, issuing battle commands the crew ignored.[21] When the *San António* got sufficiently close, an armed party from the *Trinidad* boarded and quickly took over the ship. Magellan then sent a longboat to the *Concepción*, facing no resistance from the crew.[22]

Some of the first-hand accounts for these eventful couple of days at Port San Julián read closer to a novel than to an accurate report. Even after discarding some of the more fantastic accounts—such as the *San António's* last anchor being surreptitiously cut by a covert sailor sent by Magellan[23]— what is left still paints Magellan as a stone-cold strategist who timed his attack to perfection, and the Spanish officers as frightened incompetent villains. History is too often rewritten by those that come out victorious, so it perhaps better to assume Magellan's success was due not only to good planning but also to a good dose of luck, a mixture which would come into play several more times during the voyage. As for the Spanish officers, they were certainly

---

[19] Adapted from: Tim Joyner, *Magellan* (Maine: International Marine, 1992), on p. 137–143; Samuel Eliot Morison, *The European Discovery of America - The Southern Voyages* (New York: Oxford University Press, 1974), on p. 373.

[20] Jean Denucé, *Magellan* [...], on p. 274.

[21] José Toribio Medina, *El descubrimiento del Océano Pacífico* [...], on p.CCXIX; Jean Denucé, *Magellan* [...], on p.274–275.

[22] Antonio Herrera, *Historia general* [...], on p. 299.

[23] Charles Parr cites the testimony of the *San António's* pilot (Charles Parr, *So Noble a Captain: The Life and Times of Ferdinand Magellan*, New York, Thomas Y. Crowell Company, 1953, on p. 299).

not villains. Magellan was willing to sacrifice the crew for his own personal gain, and likely many more would have survived if the mutiny attempt succeeded in bringing the ships back to Spain. Regrettably for the sailors, both rebellion attempts had been short sighted and fraught with blunders.

A court martial was set up, presided by Mesquita. With the dead body of Mendoza propped up and standing next to the other accused, the court swiftly found 40 men guilty of mutiny.[24] The stipulated penalty for such a crime aboard the king's ships was death,[25] but Magellan knew he could not kill 15 per cent of his crew. Who would carry out the horrific deed? Who would man the ships afterwards? And how would Magellan and Mesquita live with a conscience weighted down by the death of so many fellow mariners? Their compromise was to commute most sentences. In lieu of death, the mutineers would spend the winter performing the hardest, riskiest, and most unpleasant tasks required to bring the ships back to running order. Quesada however, as the leader of the uprising, got his full sentence. He was decapitated (a dubious honour reserved for nobleman, as common men were garrotted instead), quartered, and his remains put on gibbets as a grim reminder of the perils of mutiny.[26] The already dead Mendoza was subject to the same fate.

Two mutineers were spared from death and hard labour: Cartagena— Magellan's *conjunta persona*—and Bernard Calmette, the French chaplain of the *San António*. The pair would incite another rebellion over the winter, one that was quickly exposed. This time, Magellan followed through with his threats and marooned both the Spanish noblemen and the French man of the cloth.[27] The sentence was carried out a few months later, when the fleet left Port San Julián. The pair was left behind with a supply of wine and biscuit, and never heard from again.

History would repeat itself 58 years later, in Francis Drake's circumnavigation voyage. Drake's fleet, authorised by Queen Elizabeth I, was officially a discovery expedition but its covert objective was to raid Spanish convoys transporting valuable goods from the Spanish colonies. Drake decided to spend the winter at Port San Julián, where he reportedly found the gibbets

---

[24] José Toribio Medina, *El descubrimiento del Océano Pacífico* [...], on p.CCXX.

[25] Pablo Emilio Pérez-Mallaína Bueno, *Spain's Men of the Sea* (Baltimore: The Johns Hopkins University Press, 1998), on p. 208.

[26] Gaspar Correia in Charles Nowell, ed., *Magellan's Voyage Around the World: Three Contemporary Accounts* (Evanston: Northwestern University Press, 1962), on p.319.

[27] Juan Sebastián Elcano *in Información recibida por el alcalde de casa y corte, Santiago Díaz de Leguizamo, en que declaran el capitán de la nao Victoria, Juan Sebastián Elcano, con Francisco Albo y Fernando de Bustamente, sobre distintos pormenores del viaje de la primera vuelta al mundo* (Seville: Archivo General de Indias, ES.41091.AGI//PATRONATO,34,R.19, 1522), http://pares.mcu.es/Par esBusquedas20/catalogo/description/122228.

used for hanging the remains of Quesada and Mendoza. During his stay, Drake also held a court martial for mutiny. The accused Thomas Doughty, Drake's co-commander, was found guilty and given a choice between decapitation and marooning. Maybe suspecting Cartagena and Calmette had endured a slow and painful death in those bleak landscapes, Doughty chose the swifter decapitation.[28]

6.1   What Did 16th Century Sailors Eat?

The short answer

Unless the captain had ordered food rationing, sailors enjoyed three hearty meals per day, an uncommon occurrence in the 16th century. However, the menu was made of items chosen for durability rather than taste: sea biscuit (hardened bread which needed to be soaked to be edible), salted meat or fish, legumes or rice, a bit of cheese, and water and wine. Officers ate the same, but often with better ingredients, a desert, and enjoying something exceedingly rare aboard a ship: a table and chairs.

Food, or rather the lack of it, would be a growing source of conflict and misery throughout the expedition. In normal circumstances though, sailors could count on three filling meals a day, a luxury most landlubbers in sixteenth century Europe did not enjoy. Still, apart from the regularity and the adequate caloric intake, there was nothing sumptuous about a sailor's ration.[29]

Its staple was biscuit, an unleavened bread baked twice to endure many months of storage without spoiling. Biscuit was so hard it needed to be soaked for several minutes before it was edible. Its name has long lost its original meaning (from Latin, *bis* meaning twice and *coctus* meaning cooked) and is today used to refer to more appetising baked sweets.[30]

Sailors preferred to soak biscuit in wine rather than in water to make it more palatable. In fact, they drank as much wine as they did water. Not only did the alcoholic beverage make life aboard somewhat less miserable, but it also tasted—and smelled—better than stagnated water and provided much needed extra calories.[31]

---

[28] Samuel Eliot Morison, *The European Discovery of America* [...], on p. 642–643.

[29] Pablo Emilio Pérez-Mallaína Bueno, *Spain's Men of the Sea* (Baltimore: The Johns Hopkins University Press, 1998), on p. 140, 141.

[30] Idem, on p. 141.

[31] Idem, on p. 141, 142.

Table 6.1 Typical weekly menu for Spanish fleets[33]

Day of the week	Menu
Monday	690 g of biscuit 150 g of salted fish 150 g of mixed horse beans and chickpeas 1 l of water 1 l of wine
Tuesday	690 g of biscuit 230 g of salted pork 46 g of mixed rice and oil 1 l of water 1 l of wine
Wednesday	*Same as Monday*
Thursday	690 g of biscuit 460 g of salted meat 60 g of cheese 1 l of water 1 l of wine
Friday	*Same as Monday*
Saturday	*Same as Monday*
Sunday	*Same as Thursday*

Biscuit and wine were supplemented with legumes (chick-peas, beans, and lentils), salted fish or meat, and cheese (Table 6.1). The fleet had left Spain with livestock[32] so, while it lasted, the crew enjoyed a few meals of fresh meat. Every ship also carried an assortment of fishing line and hooks: if there was idle time, sailors could attempt to catch the next meal.

Daily rations varied slightly throughout the week and were divided into three meals. A light breakfast was served at sunrise, usually made up of wine and a bit of salted meat or fish. Lunch, served around noon, was the day's main meal. Dinner came before sunset, usually with half of the amount of food served at lunch.[34]

Daily rations provided an estimated 3,500 to 4,200 cal, enough for the efforts demanded from a sailor – today's recommended intake hovers around 2,500 cal for men, but our days are seldomly as frantic as a mariner's. However, cutting rations in half—something which happened often aboard

---

[32] Martin Fernandez de Navarrate, *Coleccion de los viajes y descubrimientos que hicieron por mar los españoles desde fines del siglo XV*, Tomo 4, Madrid, Imprenta Nacional, 1837, on p. 187.

[33] Example from 1560, adapted from Earl Hamilton, "Wages and Subsistence on Spanish Treasure Ships, 1503–1660", *Journal of Political Economy*, Vol. 37(4), 1929, p. 430–450, on p.434, citing documents from Casa de La Casa de la Contratación at the Archivo General de Indias.

[34] Pablo Emilio Pérez-Mallaína Bueno, *Spain's Men of the Sea* [...], on p. 143.

Magellan's fleet and other sixteenth century ships—left the crew malnourished. Even when enjoying full rations, the lack of fresh fruits and vegetables meant sailors did not ingest enough vitamins and often fell ill with scurvy. The one litre daily allowance of water was also well below the two to ten litres (in cold climates and in the tropics, respectively) that somebody doing physical work should consume. The additional allowance of one litre of wine per day was generous from an alcoholic content perspective, but not enough to make up for the lack of water.[35]

Officers had the same base diet as the crew, but often with finer ingredients (white biscuit in place of dark biscuit, better wine). They also enjoyed the flavour of a couple of condiments—capers and mustard—and a bit of a desert—raisins, currants, figs, almonds, and quince jelly.[36] Besides catering for a sweet tooth, some of these complements also have modest amounts of vitamin C, perhaps enough to (unknowingly) delay the effects of scurvy.[37]

Officers ate separately from the crew and were the only ones who enjoyed something exceptionally rare in sixteenth century ships: a table and chairs. They were also the only ones who had servants cooking for them. The rest of the mariners was expected to cook their own food, which was seen as a demeaning task (*"your beard reeks of cooking"* was a grave insult). To avoid it, many paid an apprentice or joined a rancho. Each rancho prepared a communal bowl of food, propped up their sea chests into a makeshift table, and ate together. There was not enough plates and utensils to go around, so sailors helped themselves directly from the bowl using their hands and the knives they always carried on their belts.[38]

6.2   How Did 16th Century Exploration Change Food Habits Around the Globe?

The short answer

The arrival of the Europeans to the Americas interrupted some 20,000 years of isolation, which had been in place since Beringia—a land bridge between Asia and the Americas—disappeared. During this time, the food crops on both sides of the Atlantic diverged substantially. The exchange that followed brought potatoes, sweet potatoes, cassava and other New World foods to the Old

---

[35] Pablo Emilio Pérez-Mallaína Bueno, *Spain's Men of the Sea* [...], on p. 143, 144.

[36] Idem, on p. 142, 143.

[37] See Crossing the world's largest ocean.

[38] Pablo Emilio Pérez-Mallaína Bueno, *Spain's Men of the Sea* [...], on p.142, 143.

World (Europe, Asia, and Africa), where they complemented local crops rather than competing for the same land resources. On the other hand, sugar, coffee, bananas and other Old World foods thrived in the New World, benefiting from favourable climates and the lack of local parasites and pests. Not all exchanges were beneficial though: tobacco and cocaine, traditionally used by Native Americans for medicinal and religious purposes, took the Old World by storm.

As the fleet progressed and the original provisions were depleted, the crew adapted their diet to the available local produce. Similarly, other European expeditions not only subsisted on American species but also brought them back to Europe. Columbus, for instance, offered sweet potatoes as proof he had reached land.[39] In parallel, Europeans introduced species in the Americas. Over time, this two-way exchange dramatically changed food habits not only in the Americas and Europe, but across the entire planet.

The sweet potatoes brought back by Columbus—and savoured by Pigafetta in Brazil—are today a staple of diets throughout the South Pacific (Solomon Islands, Timor-Leste), Asia (China) and Eastern Africa (Rwanda, Burundi, Uganda). On a *per capita* basis, the only two American countries remaining in the top 10 of sweet potato consumption are Cuba and Haiti (Fig. 6.3). Similarly, cassava—which Pigafetta would later note to be consumed in vast quantities in Patagonia—is today mostly associated with African cuisine. The rapid adoption of these foods in the Old World –across Europe, Asia, and Africa—was caused by how well they complemented local crops. Rather than competing for the same land, these new plants would often grow in different seasons or make use of soils previously deemed useless for agriculture.[40]

In Europe, the potato was the most revolutionary plant brought from the Americas. Endemic to the Andes, where they were probably domesticated 7,000 to 10,000 years ago,[41] potatoes are notoriously unfussy and able to grow across most latitudes. These tubers provide ample calories and most of the base nutrients required for a sustainable diet: humans can survive only on potatoes and a complement of milk, which delivers the missing vitamins A and D. Initially Europeans looked at potatoes with suspicion, as they grew from tubers and not from the more familiar seeds. Some feared

---

[39] Patricia O'Brien, "The Sweet Potato: Its Origin and Dispersal", *American Anthropologist*, 74(3), 1972, p.342–365, on p.345.

[40] Alfred Crosby, *The Columbian Exchange: Biological and Cultural Consequences of 1492* (Westport: Praeger Publishers, 2003), on p.177.

[41] Ellen Messer, "Maize," *The Cambridge History of World Food*, edited by Kenneth F. Kiple and Kriemhild Coneè Ornelas, Volume I (New York: Cambridge University Press, 2000), on p. 97–112.

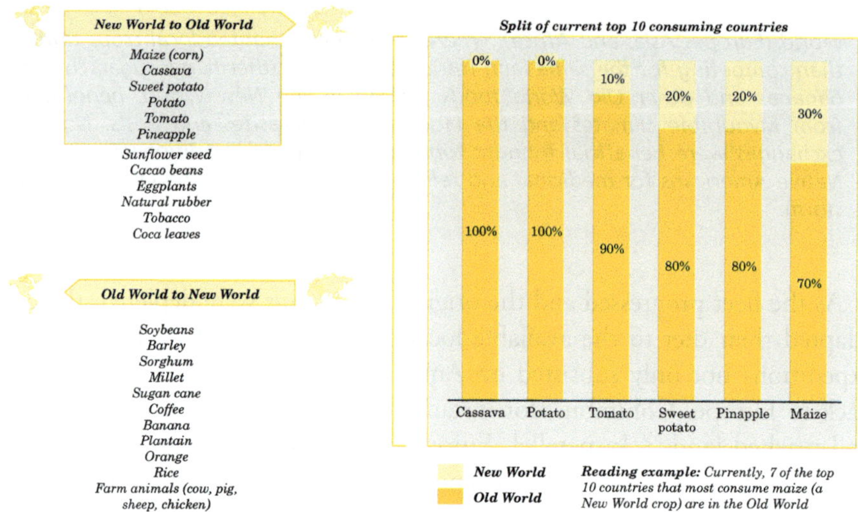

Fig. 6.3 Overview of food exchanges between the Old and New Worlds[42]

they could be poisonous due to its similarity to nightshades.[43] Others were disgusted by their discoloured and lumpy appearance, which they associated with leprosy. Over time, however, the potato's benefits outweighed its strangeness. By mid-eighteenth century, the vegetable had made its way across the entire continent, from the Iberian Peninsula to Scandinavia. Its impact on nutrition and health was astonishing: estimates suggest that the introduction of potatoes accounted for at least 25 per cent of Europe's population increase in the eighteenth and nineteenth centuries. They also had a similar impact on urbanisation, a key precursor for the Industrial Revolution.[44]

Pineapples were not quite as revolutionary as potatoes, but they also made quite an impression in the Old World. Pigafetta could not have enough of them in Brazil, describing them as *"sweet pine cones, in truth the most delicious fruit that can be found"*.[45] Descriptions of pineapples had reached Europe

---

[42] Adapted from: Nathan Nunn and Nancy Qian, "The Columbian Exchange: A History of Disease, Food, and Ideas", *Journal of Economic Perspectives*, 24(2), 2010, p. 163–188, on p.168 (Table 6.1), p.170 (Table 2), doi.org/10.1257/jep.24.2.163; Simon Lewis and Mark Maslin, "A transparent framework for defining the Anthropocene Epoch", *The Anthropocene Review*, 2(2), 2015, on p.134, Fig. 6.2, doi.org/10.1177/2053019615588792.

[43] Charles Brown, "Origin and history of the potato", *American Journal of Potato Research*, 70, 1993, p.363–373, on p. 364.

[44] Nathan Nunn and Nancy Qian, "The Potato's Contribution to Population and Urbanization: Evidence from an Historical Experiment", *The Quarterly Journal of Economics*, 126(2), 2011, p.593–650, doi.org/10.1093/qje/qjr009, on p. 593.

[45] Antonio Pigafetta, *The first voyage around the world 1519–1522*, edited by Theodore Cachey Jr. (Toronto: University of Toronto Press, 2007), on p. 8.

before—starting with the return of Columbus' second expedition—but the fruit itself rarely survived the voyage.[46] It eventually made its way across the Atlantic, first as a sugar syrup preparation and then in its full glory, after being transplanted and acclimatised to suitable eastern climates. Today, pineapples are a national dish across the entire tropical belt, from Costa Rica to Samoa. Its name, however, hints at two separate distribution paths. In Spanish it is called *piña* (pinecone), alluding to the resemblance noted by Pigafetta and others. English adopted a similar logic (in Old English, "apple" was a generic term for fruit), and the name followed the introduction of the fruit in countries such as Japan (*painappuru*) and Korea (*pain-aepeul*). In the meantime, a French priest travelling through Brazil noticed Tupi natives called the fruit *nana* ("excellent fruit") the precursor for the *ananas* denomination that was introduced to most other languages.[47] While this is also the case in European Portuguese, Brazilian Portuguese adopted instead *abacaxi* (derived from "fragrant fruit" in Tupi).[48]

Not all plants brought from the Americas had a positive impact in the Old World, though. Tobacco was consumed by native Americans since the first century BE, but seemingly only during religious ceremonies and as a painkiller. It quickly spread throughout Europe during the sixteenth century, most notably by the hands of Jean Nicot de Villemain, the French ambassador in Portugal (from whom "nicotine" and tobacco's binomial name—*Nicotiana tabacum*—originate). Nicot advocated the use of tobacco for medicinal purposes – arguing for instance that it was an effective deterrent against the Black Plague—but it was soon also used recreationally. In the twentieth century the use of tobacco skyrocketed till the 1960s, when a report from the U.S. Surgeon General linked smoking to lung cancer and other severe health issues.[49] Decades of antismoking campaigns brought worldwide tobacco consumption down, but today it is still consumed by more than 20 per cent of the global population and kills more than 8 million per year.[50]

Just like tobacco, native Americans had been consuming coca leaves for religious and medicinal purposes well before the arrival of Europeans. Spanish

---

[46] Teresa Carvalho, "The Natural Frontiers of a Global Empire: The Pineapple—Ananas comosus—In Portuguese Sources of the 16th Century", *Humanities*, 9(3), 2020, p.1–21, doi.org/10.3390/h9030089, on p. 4.

[47] Arabic, Dutch, German, Greek, Hindi, Norwegian, Polish, Turkish and Russian, just to name a few.

[48] Teresa Carvalho, "The Natural Frontiers of a Global Empire" […], on p. 15.

[49] Nathan Nunn and Nancy Qian, "The Columbian Exchange: A History of Disease, Food, and Ideas" […], on p. 175, 176.

[50] "Tobacco fact-sheet", World Health Organization, 2023, https://www.who.int/news-room/fact-sheets/detail/tobacco.

settlers noted that chewed coca leaves acted as a stimulant and could help suppress hunger, thirst, pain, and fatigue. The popularity of the natural medicine grew across both the New and the Old Worlds but remained relatively benign until the mid-nineteenth century, when the cocaine alkaloid—the active ingredient of coca leaves—was isolated. On both sides of the Atlantic, many physicians exulted the qualities of this new miracle drug in the treatment of a flurry of health and mental issues. However, cocaine—which is about one hundred times more potent than coca leaves—soon showed its dark side. Physicians such as Sigmund Freud (an Austrian neurologist and psychoanalyst) and William Halsted (a US surgeon), who had been early advocates of cocaine, battled with addiction for decades.[51] Cocaine was first made illegal in the US in 1914, and other countries soon followed suit. Today, cocaine and other illicit drugs kill over half a million users per year. South America—Peru, Colombia, and Bolivia[52]—continues to be the prime producer of coca leaves, but the resulting cocaine is overwhelmingly consumed in the US (~65%) and Europe (~30%).[53]

Much like American plants flourished elsewhere, many Old World crops also found suitable lands in the New World. Several of them—including sugar cane, coffee, bananas, oranges, soybeans, barley, and millet—are today disproportionally grown in the Americas, despite being endemic to other regions. A couple of reasons help to explain this: first, the Americas cover most of the latitude range, making it easier to find a climate suiting each transplanted crop; second, the insularity of the Americas meant local pests and parasites had not evolved to attack foreign crops. Tropical crops such as coffee beans and sugarcane benefited the most from the lack of parasites and pests, striving in the Americas.[54] Today, Brazil is the major producer of these two plants, by a good margin.[55]

Together with these crops, Europeans also brought an invisible enemy that would soon change American demographics forever: a roster of diseases for which the natives had no immunity against.[56]

---

[51] Nathan Nunn and Nancy Qian, "The Columbian Exchange: A History of Disease, Food, and Ideas" […], on p. 175–177.

[52] Idem, on p.177.

[53] "Estimating illicit financial flows resulting from drug trafficking and other transnational organized crimes", United Nations Office on Drugs and Crimes, 2011, https://www.unodc.org/documents/data-and-analysis/Studies/Illicit_financial_flows_2011_web.pdf, on p.58 (Table 39).

[54] Nathan Nunn and Nancy Qian, "The Columbian Exchange: A History of Disease, Food, and Ideas" […], on p.177–179.

[55] In 2025, Brazil produced nearly 35% of the world's coffee ("Production – Coffee", USDA, 2025, https://fas.usda.gov/data/production/commodity/0711100) and nearly 25% of the sugar ("Production – Sugar", USDA, 2025, https://fas.usda.gov/data/production/commodity/0612000).

[56] See What Happened to the Natives Encountered by the Fleet?

7

Winter in Patagonia

From Port San Julián (April 3rd, 1520) to Santa Cruz (October 18th, 1520)

With the uprising quelled, it was now time to carry on with the ship repairs needed after the hazardous exploration of the South American coastline. After more than seven months of sailing, the ships also needed a good scrubbing from mast to keel.[1] The punished mutineers took on the hardest tasks, but there were plenty of unpleasantries to go around (Fig. 7.1).

A camp was set up ashore, with a stone smithy for building replacement iron parts plus sheds to store goods and provisions while the ships were repaired and cleaned. The stones used for ballast, foul-smelling from months of simmering in salt water and filth which had made its way down to the bottom of the vessels, were taken out and left in the shoreline for the waves to clean. Once empty, the ships were careened so the hull could be repaired. Rotten and damaged planks were replaced, the seams were caulked, and the portion of the hulls below the waterline were tarred. Above deck, the ripped sails were sewn and the failing rigging overhauled.[2] The most complex repairs were well beyond the skills of most sailors, who left them for the caulkers and carpenters. It was now clear why these craftsmen were taken on board and earned more than the regular sailors.[3]

Magellan had initially assumed there would be plenty of time during the winter to conduct the repairs, but he learned otherwise. When the provisions were inventoried, they found only half of what they were expecting.

---

[1] Tim Joyner, *Magellan* (Maine: International Marine, 1992), on p. 144.

[2] Idem.

[3] See What Was the Life of a 16th Century Sailor Like?

© The Author(s), under exclusive license to Springer Nature
Switzerland AG 2026
R. Gaspar, *The Revolution of Magellan*, https://doi.org/10.1007/978-3-032-10797-8_7

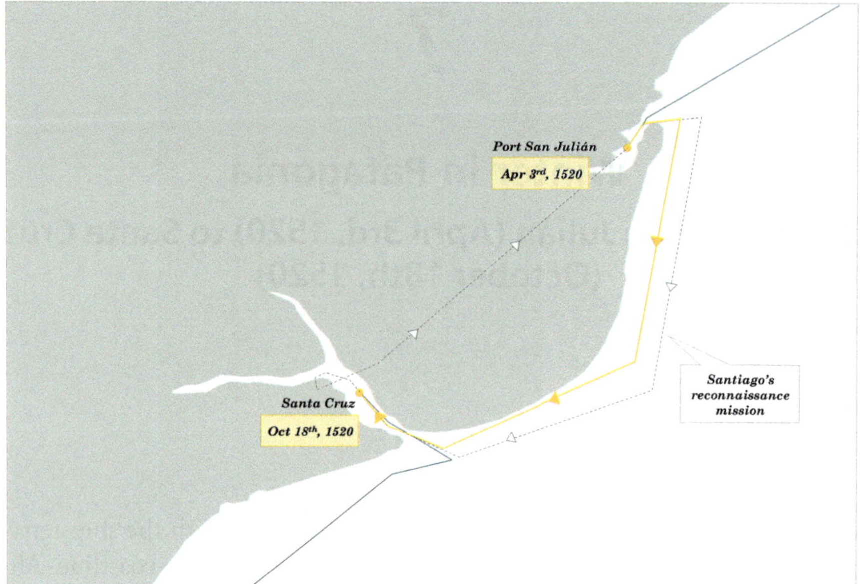

Fig. 7.1 Fleet route[4]

The embezzlement had gone unnoticed so far because the culprits had issued two invoices for each delivery. The remaining provisions would only last for about six months[5] and the local game would hardly make up for the difference. While the crew could see plenty of large rodents and guanacos (wild llamas, at the time unfamiliar to Europeans) roaming around, they did not know how to hunt them. Fresh water sources were also scarcer than initially assumed.[6]

And the list of hurdles went on. Not even the hasty pace Magellan imposed to the repair operations was enough to keep the crew warm, and several men died of cold or suffered severe frost bite.[7] While there are plenty of accounts corroborating the cold-induced misery, it is nevertheless surprising it got to these extremes, as average low temperatures in Port San Julián usually hover

---

[4] Fleet route adapted from Tomás Mazón Serrano ("Mapas", Ruta Elcano, June 19, 2024, https://rut aelcano.com/mapas/); map adapted from Natural Earth (1:10 m large scale land data, version 5.1.1, https://www.naturalearthdata.com/downloads/).

[5] Charles McKew Parr, *So noble a captain: The life and times of Ferdinand Magellan* (New York: Crowell, 1953), on p. 302, 303.

[6] Tim Joyner, *Magellan* […], on p. 144.

[7] Antonio Herrera, *Historia general de los hechos de los castellanos en las islas i Tierra Firme del Mar Oceano* (Madrid: En la emplenta Real, 1601), Decada II, Libro IX, Cap. XIII, on p. 299, 300.

around 0° C in winter,[8] typical for the 49° parallel. The 1520 winter may have been particularly cold,[9] catching the fleet—departed from sunny Seville and perhaps not expecting to go much below the 35° parallel—off guard.

By early May. the small and nimble *Santiago* was already repaired, so Magellan sent it south on a reconnaissance mission.[10] After a few days tacking against the strong head wind, the *Santiago* spotted an opening in the shoreline. Could it be the eagerly sought passage to the west, at little more than 100 km from Port San Julián? Not likely, as the opening was narrow, but it could still prove to be a good food source. Sound in hand, navigating cautiously around the shoals, Serrano and his crew went in and followed the water way as it curved northwest. Shortly after, they found themselves at an intersection, close to a sheltered harbour Serrano named Santa Cruz. Spotting many seals, they spent a couple of weeks slaughtering them and smoking the meat, amassing provisions for their companions.[11]

They then resumed the reconnaissance mission, deciding to explore first the larger opening to the west. It was a short trip, as an unexpected squall tore away their sails and rudder. Without propulsion and steering, large winter waves sent the *Santiago* crashing onto the shore. Except for an unfortunate sailor, the entire crew reached land before the surf tore the vessel down to pieces. All their provisions, including the smoked seal meat they had laboured over for two weeks, were lost.[12]

The survival plan outlined by the castaways was a complicated one: collect planks from the wreck while subsisting on the local barnacles, walk back to the Santa Cruz harbour, build a raft to cross to the northern shore, and then hike back to Port San Julián. Weakened by cold and hunger, they discarded most of the planks before reaching Santa Cruz, arriving with only enough wood to build a two-men raft. Serrano selected his two strongest and youngest sailors to cross the estuary in the raft, while he stayed back with the rest of the mariners. The distance from Santa Cruz to Port San Julián is about 80 km in a straight line, but the two young mariners surely walked more than that, as they had to negotiate their way around rough terrain. After nearly two

---

[8] "Provincia de Santa Cruz - Clima Y Meteorologia: Datos Meteorologicos Y Pluviometicos", Secretaria de Mineria, retrieved 2015, https://web.archive.org/web/20150119032810/http://www.mineria.gob.ar/estudios/irn/santacruz/tablametypluvio.asp; "Estadísticas Climatológicas Normales - período 1991–2020", Servicio Meteorológico Nacional, retrieved 2024, https://repositorio.smn.gob.ar/handle/20.500.12160/2506.

[9] Meteorological data for Port San Julián, collected since 1951, shows a record low of -14° C.

[10] Antonio Herrera, *Historia general* [...], on p.299, 300.

[11] José Toribio Medina, *El descubrimiento del Océano Pacífico: Vasco Núñez de Balboa, Fernando de Magallanes y sus compañeros* (Santiago de Chile: Imprenta Universitaria, 1920), on p.CCXXVII.

[12] Idem.

weeks they finally reached their destination, so thin and weak their comrades did not recognise them at first.[13]

After learning what happened, Magellan did not seem to hesitate to send a rescue party to get back the stranded men, which was not something to take for granted in the sixteenth century. In fact, some 10 years before, while aboard a Portuguese ship in the Indian Ocean, Magellan had protested when his commander wanted to leave a crew of 28 behind. The Portuguese ship had raided a Malay junk packed with goods and was towing it back to port. Unnoticed by the boarding party, the Malay crew had split the hull. The Portuguese commander (Diogo Lopes de Sequeira, who would later become Governor of Portuguese India), fearful the sinking junk would drag his ship underwater, ordered the towing rope to the cut. Magellan, seeing the boarding party abandoned to their fate, cried out *there could do no better feat than to save our men on that junk!*. Sequeira, while irritated by the challenge, conceded and sent Magellan leading a rescue party which succeeded in saving all Portuguese aboard the junk (the Malay crew, who had been confined below deck, was apparently left behind).[14]

Now, even if Magellan was as cold-hearted as Sequeira, it would have been foolhardy to leave his men stranded in Santa Cruz, as it would take away the last remnants of morale from his crew. The wintery conditions would put another ship at risk, so Magellan sent instead a rescue party of 20 men overland, retracing the steps made by the two young sailors. Little is known about the rescue, other than it successfully brought back the *Santiago* crew by late June, almost three months after they had left Port San Julián.[15] The *Santiago*'s reconnaissance mission had cost the fleet a ship, but it had not been a complete wash, as they knew that Santa Cruz could provide them with much needed supplies.

During the *Santiago* ordeal, a lot had also happened at Port San Julián. The crew had previously noticed some very large footprints in the sand but had not yet seen anyone in those barren landscapes. In early June, they finally spotted a man on the shoreline. He was singing, dancing, and throwing sand over his head. Magellan, eager to get a closer look, ordered a mariner to approach him and mimic his movements to establish trust. It worked and the local came onboard, where he was showered with gifts and food.[16]

---

[13] José Toribio Medina, *El descubrimiento del Océano Pacífico* [...], on p.CCXXVIII.

[14] Visconde de Lagôa, *Fernão de Magalhães: A sua vida e a sua viagem* (Lisboa, Seara Nova, 1938), on Volume I, p.132, citing Castanheda.

[15] José Toribio Medina, *El descubrimiento del Océano Pacífico* [...], on p. CCXXVIII.

[16] Idem, on p. CCXXVIII, CCXXIX.

Over the next weeks, the crew met with more than 20 natives, who they called Patagonians.[17] Pigafetta's diary once again came alive with descriptions of the indigenous and their customs. About that first encounter, the Italian reported "he was *so tall that we reached only to his waist, and he was well proportioned*". The women "*are not so tall as the men but are very much fatter. When we saw them we were greatly surprised: their breasts are one-half cubit long and they are painted and clothed like their husbands, except that in front of their private parts they have a small skin that covers them*". Pigafetta was also impressed by how they hunted the guanacos, something which had eluded the crew: "*When those people wish to take some of those animals, they tie one of these young ones to a thornbush; thereupon, the large ones come to play with the little ones, and those people kill them with their arrows from their place of hiding*".[18]

These "Patagonians" were likely Tehuelche.[19] Pigafetta's exaggerated description of their height and feet may have been the origin of the myth of the Patagonian giants, which endured in Europe for more than 250 years. Later accounts estimated their height at 180 or 190 cm,[20] still taller than typical sixteenth century Southern Europeans[21] and certainly towering over Magellan, who was apparently quite short.[22]

The trust placed by the Tehuelche on the Europeans would soon prove ill-fated. Eager to bring a few natives back to Spain, the crew set their eyes on a couple of young and strong specimens. To avoid a fight, they first filled their hands with presents. The last present was a pair of iron fetters they could no longer carry, so the crew offered to place them around their ankles. When the gullible duo realised the trick, the shackles had already been riveted in place. Magellan then ordered ashore a crew of armoured men, apparently to

---

[17] The origin of the name remains disputed: some consider it a contraction of *pata de cano*—dog paw—owning to the large footprints their hide-covered feet left on the sand; while others note that *Primaleòn*, a well-known contemporary Spanish poem, had a rather monstrous character named *Patagone* (Antonio Pigafetta, *The first voyage around the world 1519–1522*, edited by Theodore Cachey Jr., Toronto, University of Toronto Press, 2007, on p. 143).

[18] Antonio Pigafetta, *The first voyage around the world 1519–1522* […], on p. 12, 13.

[19] See What Happened to the Natives Encountered by the Fleet?

[20] José Toribio Medina, *El descubrimiento del Océano Pacífico* […], on p.CCXXIX, cited Darwin's estimate of 183 cm and Muster's estimate of 193 cm.

[21] Contrary to popular belief, medieval Europeans were not significantly shorter than today. For instance, an analysis based on archaeological excavations suggests than the average height of Englishmen from 1400 to 1650 was 173–174 cm, compared to 177 cm in 1970 (Gregori Galofré-Vilà, "Heights across the Last 2,000 Years in England", *Research in Economic History*, 34, 2017, p. 1–36, on p. 14).

[22] José Toribio Medina, *El descubrimiento del Océano Pacífico* […], on p.CVI, cited Las Casas' encounter with Magellan.

capture the wife of one of the captives, but the squad came empty-handed and missing one sailor, killed in a skirmish.[23]

On August 24, with the repairs complete and taking advantage of an opening in the weather, the fleet left Port San Julián headed to Santa Cruz, where they would stay for two months replenishing supplies. They took the two Tehuelche captives with them but left behind Cartagena and Calmette, the leaders of the last failed mutiny attempt.[24] After what the crew had done to the locals, the prospects for the Spanish nobleman and the French chaplain did not look good.

7.1   What Happened to the Natives Encountered by the Fleet?

The short answer

Today, there are less than 20,000 descendants of the tribes encountered by Magellan's fleet—the Tamoios, the Querandí, and the Tehuelches. Across the Americas, the native population is thought to have decreased by some 70% after the arrival of the Europeans, and more than 12 million slaves were brought from Africa to make up for the dwindling local labour force. War, mistreatment, and negligence contributed to the sharp decrease in population, but not as much as disease. European ships carried a suite of Old World maladies to which the natives had never been exposed to, including smallpox, measles, bubonic plague, whooping cough, typhus, and malaria. Smallpox was the deadliest, as it killed about 50% of the infected, and each patient infected five more. The variola virus may well have been humanity's worst enemy so far, as it killed an estimated 300 to 500 million people worldwide.

In the more than 10 months it took the expedition to sail down from Rio de Janeiro to Santa Cruz, Magellan and his men met only a handful of native peoples (Fig. 7.2). Many others were perhaps hunting further inland or cautiously hiding away from the unfamiliar visitors. In due time though, apart from a few tribes which continue to live isolated deep in the Amazon rainforest, they would all face the Europeans.

Back at Rio de Janeiro, the amicable relations[25] between the Tamoios and the Portuguese did not last long. By the 1550s, the two parties found themselves in opposite sides of a war. On one side stood the Tamoios

---

[23] Antonio Pigafetta, *The first voyage around the world 1519–1522* […], on p. 14, 15.

[24] José Toribio Medina, *El descubrimiento del Océano Pacífico* […], on p.CCXXXI.

[25] See Out of Portuguese Territories.

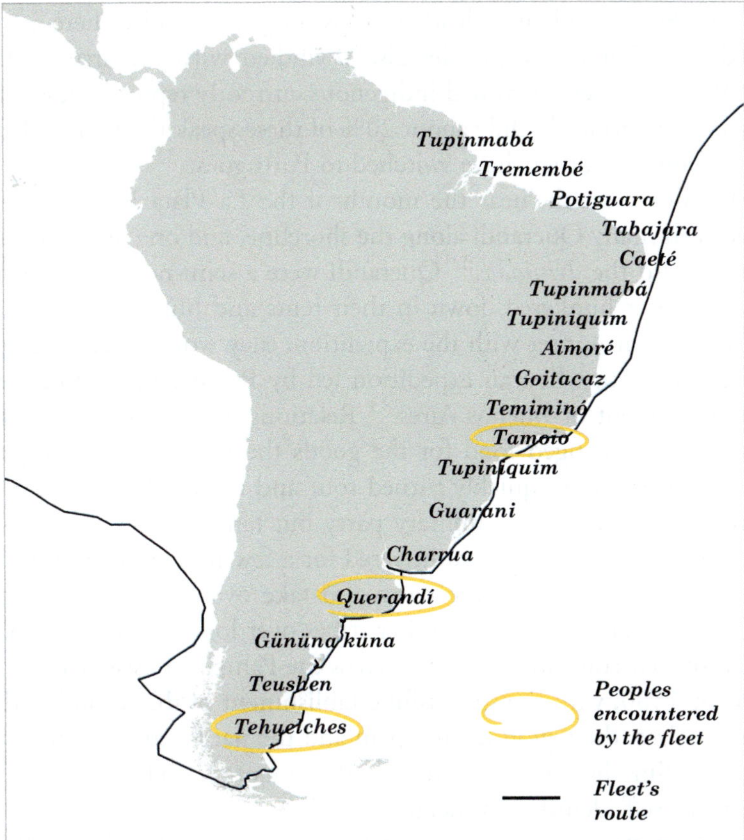

Fig. 7.2 Major indigenous peoples that inhabited the eastern South American coastline in the sixteenth century, most of them not encountered by the fleet[26]

and the French, who had also established a colony in the region. On the other faction, the Portuguese allied with the Tupiniquins and other rival tribes. The war dragged on for over a decade but finally ended in 1567, when Portuguese reinforcements ousted the French and hunted down the Tamoios.[27] The once numerous people, estimated to have been around 70,000 before European contact,[28] became virtually extinct. Today, little

---

[26] Adapted from: Eduardo Bueno (*Capitães do Brasil* [...]) and Willem Adelaar (*The Languages of the Andes*, Cambridge, Cambridge University Press, 2004); fleet route adapted from Tomás Mazón Serrano ("Mapas", Ruta Elcano, June 19, 2024, https://rutaelcano.com/mapas/); map adapted from Natural Earth (1:10 m large scale land data, version 5.1.1, https://www.naturalearthdata.com/downloads/).

[27] Aylton Quintiliano, *A guerra dos Tamoios* (Rio de Janeiro: Relume Dumará, 2003), on p. 223.

[28] Eduardo Bueno, *Capitães do Brasil – A Saga dos Primeiros Colonizadores* (Rio de Janeiro: Objetiva, 1998), on p. 66.

more than 80 individuals self-identify as Tamoios.[29] Elsewhere in Brazil, hundreds of other native peoples also dwindled with the arrival of Europeans. Altogether, self-identified indigenous currently represent less than 1% of Brazil's population.[30] Only about 20% of these speak their native language at home, while the others have switched to Portuguese.[31]

Further down south, near the mouth of the La Plata River, the expedition had seen many Querandí along the shoreline, and one of them had even climbed aboard the *Trinidad*.[32] Querandí were a semi-nomadic people who spent the winter hunkered down in their tents and hunted in the summer. After the brief encounter with the expedition, they would again engage with Spaniards in 1536, when an expedition led by Pedro de Mendoza founded the first settlement of Buenos Aires.[33] Relations were initially friendly, and the Querandí exchanged food for the goods the settlers had brought from Spain. However, things quickly turned sour and the locals stopped bringing food. Mendonza sent out a military party but failed to force the Querandí into submission. The settlement lingered for a few more years but was abandoned in 1542, prompting the natives to take over the horses left behind to better hunt game, trade with other tribes over longer distances, and fight the Spanish that continued to settle across the Pampas. Resistance was futile, though. By 1580, with the successful establishment of the second settlement of Buenos Aires, the region fell to Spain and the native culture and language were lost to time before being properly documented. Today, less than 1,000 Argentinians self-identify as Querandí.[34]

Apart from the two young men taken captive by Magellan at Port San Julián,[35] the Tehuelche (or Aónikenk, as they called themselves) enjoyed limited early contact with Europeans, particularly those in Southern Patagonia. This region had its first governor in 1529, but neither him nor his successors were in a hurry to explore and settle the barren and cold landscapes. Things started to change in the eighteenth century, with several expeditions exploring the depths of Patagonia and its peoples. Under the

---

[29] IBGE, *2010 Census*, 2010, https://www.ibge.gov.br/estatisticas/sociais/populacao/9662-censo-demografico-2010.html?edicao=9677&t=resultados.

[30] IBGE, *2022 Census*, 2022, https://www.ibge.gov.br/estatisticas/sociais/populacao/22827-censo-demografico-2022.html?edicao=38698&t=resultados.

[31] IBGE, *2010 Census* […].

[32] See Sweet Rather Than Salty.

[33] Ulrich Schmídel, *Viaje al Río de la Plata (1534–1554)* (Buenos Aires, Cabaut y Cía., 1903), on p. 45.

[34] INDEC, *Censo 2022 – Población indígena o descendiente de pueblos indígenas u originarios*, 2022, https://www.indec.gob.ar/indec/web/Nivel4-Tema-2-41-165.

[35] See Winter in Patagonia.

guise of science, several Tehuelche were abducted and brought to Europe, to be studied and exhibited in European capitals.[36] Patagonia was only fully colonised in the mid-nineteenth century, already by the freshly independent Argentinean and Chilean republics. Today, there are about 17,400 self-identified Tehuelches.[37] The last fluent speaker of Aonekko 'a'ien, their language, died in 2019.[38]

What happened in Brazil and Argentina were not isolated cases. Across the entire Americas only about a quarter of the current population has native ancestry, while the Old World, particularly Europe and Africa, make up the rest (Table 7.1).

Perhaps the original native population was small, so their DNA was swiftly diluted by that of the arriving Europeans and Africans? Most evidence points to the contrary. Large swathes of the Americas, particularly those closer to the equator, provide all the right conditions for thriving human development. In regions such as Mexico, Mesoamerica and the Andres, the European explorers encountered large advanced agricultural civilizations. Ample genetic and archaeological findings also point to a numerous pre-colonial native population. However, nobody is quite sure how numerous. The challenges of backtracking 600 years of population counts have produced a wide range of pre-Columbian population estimates, from ~8 to 112 million (Fig. 7.3).

Despite this uncertainty, there is consensus that the Native American population decreased sharply after the arrival of Europeans, with most studies putting the depopulation rate above the 50% death toll bubonic plague had in fourteenth century Europe. The higher estimates suggest a population decrease of over 95%, nothing short of an extinction event (Fig. 7.4).

There is also lively debate over what caused the native population to sharply decrease. Europeans were undoubtedly involved, but how? One possibility is they purposefully killed indigenous peoples, in a genocide of continental proportions. While there no shortage of examples of Europeans causing death and suffering, there is no evidence of a concerted effort amongst Spain, Portugal, England, France, and other European powers to eradicate American natives. Explorers and settlers regarded the New World

---

[36] Christian Báez and Peter Mason, *Zoológicos Humanos* (Santiago de Chile: Pehuén, 2006), on p. 37, 38.

[37] All in Argentina (INDEC, *Censo 2022* […]). In Chile, Tehuelches were likely assimilated by the Mapuches, which represent about 80% of the indigenous population (INE, *Censo 2017 – Población que se considera perteneciente a un pueblo originario por nombre del pueblo originario*, 2017, https://www.ine.gob.cl/estadisticas/sociales/censos-de-poblacion-y-vivienda/censo-de-poblacion-y-vivienda).

[38] Javier Domingo, "La imborrable obra de Dora Manchado: ¿la última guardiana de la lengua tehuelche?", Infobae, 2019, https://www.infobae.com/cultura/2019/01/30/la-imborrable-obra-de-dora-manchado-la-ultima-guardiana-de-la-lengua-tehuelche/.

Table 7.1 Estimated proportion of the population of American countries in 2000 that were descendants of individuals living in the Americas, Europe, Africa, and others in 1500 (reading example: in 2000, 9% of the Brazilian population descended from individuals who lived in the Americas in 1500)[39]

Country	Native American	European	African	Other	% of American population
Haiti	0%	3%	98%	0%	1.1%
Jamaica	0%	8%	89%	3%	0.3%
Trinidad and Tobago	0%	7%	46%	47%	0.1%
Cuba	3%	63%	34%	1%	1.1%
Canada	3%	85%	2%	10%	3.7%
Dominican Republic	4%	52%	44%	0%	1.1%
Uruguay	4%	92%	4%	0%	0.3%
Argentina	5%	90%	2%	3%	4.4%
Guyana	5%	1%	39%	55%	0.1%
Brazil	9%	74%	16%	1%	20.8%
United States	10%	76%	10%	5%	32.7%
Puerto Rico	18%	66%	16%	0%	0.3%
Costa Rica	30%	60%	9%	1%	0.5%
Venezuela	31%	55%	14%	0%	2.8%
Panama	36%	45%	13%	6%	0.4%
Colombia	37%	46%	17%	0%	5.0%
Chile	37%	59%	1%	2%	1.9%
Belize	39%	40%	17%	4%	0.0%
Nicaragua	40%	51%	9%	0%	0.7%
Paraguay	46%	53%	1%	0%	0.7%
El Salvador	50%	50%	0%	0%	0.6%
Honduras	52%	46%	2%	0%	1.0%
Ecuador	61%	32%	7%	0%	1.7%
Mexico	63%	30%	7%	1%	12.4%
Peru	64%	28%	6%	2%	3.3%
Bolivia	72%	27%	1%	0%	1.2%
Guatemala	74%	22%	4%	0%	1.7%
Total Americas	**24%**	**62%**	**12%**	**3%**	**100%**

as an immense commercial opportunity which required an equally vast work force. The first peoples were the most obvious source of labour, either voluntary or involuntary. It would have been illogical to eliminate the cheap local

---

[39] Adapted from: Louis Putterman and David N. Weil, "Post-1500 Population Flows and The Long-Run Determinants of Economic Growth and Inequality", *The Quarterly Journal of Economics*, 125(4), November 2010, p.1627–1682, doi.org/10.1162/qjec.2010.125.4.1627; United Nations, *World Population Prospects 2022*, 2022, https://population.un.org/wpp/.

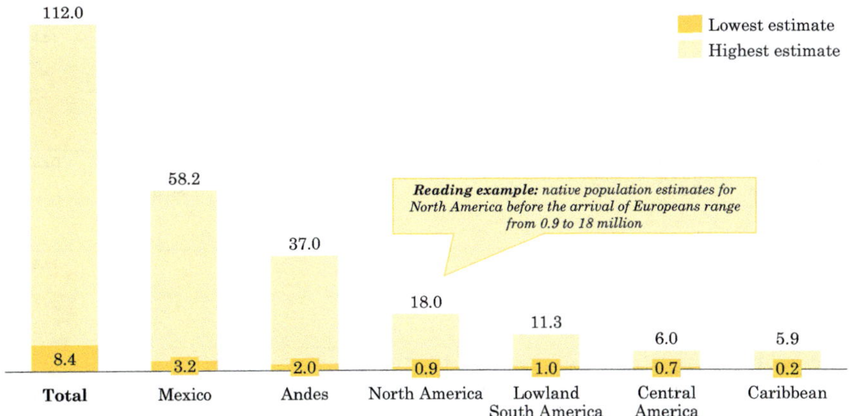

Fig. 7.3 Range of pre-Columbian American population estimates[40]

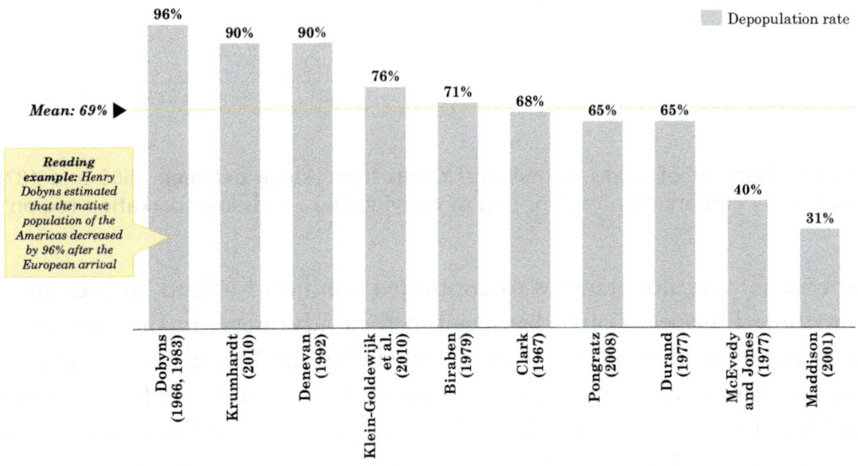

Fig. 7.4 Range of American depopulation rates after European contact[41]

resources and then incur the expense of bringing more than 12 million slaves from Africa (Fig. 7.5).

Another possibility is that the ranks and files of natives fell in the wars against the invaders. While these certainly played a role, it seems doubtful they can fully explain the entire population decline. At contact, Native

---

[40] Suzanne Austin Alchon, *A Pest in the Land: New World Epidemics in a Global Perspective* (Albuquerque: University of New Mexico Press, 2003), on p. 147–172.

[41] Adapted from Alexander Koch et al., "Earth system impacts of the European arrival and Great Dying in the Americas after 1492", *Quaternary Science Reviews*, 207, 2019, p. 13–36, https://doi.org/10.1016/j.quascirev.2018.12.004, on p. 21, Table 3.

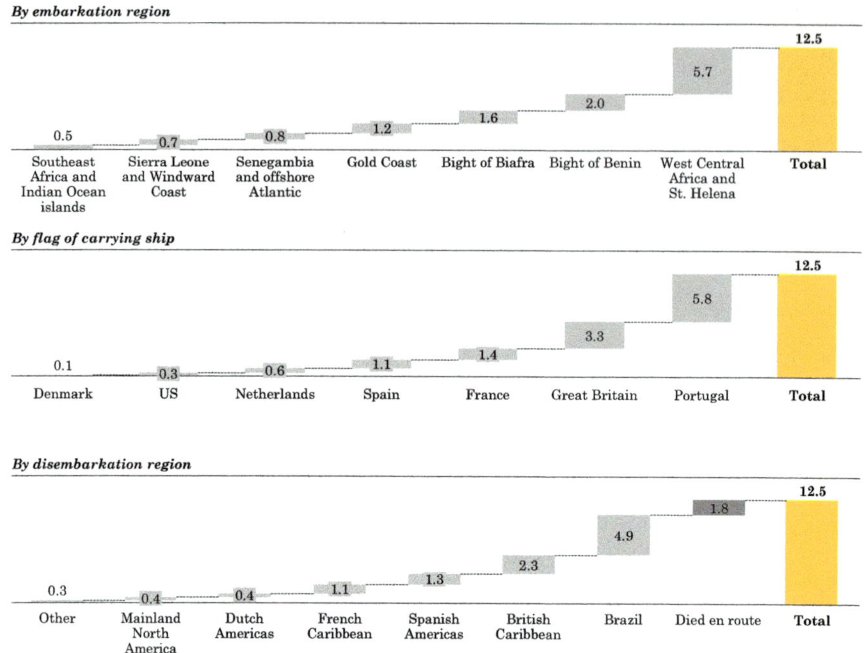

Fig. 7.5 Number of slaves (in millions) taken from Africa between 1500 and 1870, grouped by embarkation region, flag of carrying ship, and disembarkation region[42]

Americans were no strangers to continued conflicts for land and resources, something Europeans often leveraged to their advantage. In the conquest of Mexico, for instance, Hernán Cortés' allied with the Tlaxcala and other long-time enemies of the Aztec Empire, a factor which was decisive for his victory. As Magellan himself would learn,[43] a large part of the European military superiority vanished once they set foot outside a ship. The remaining guns and armour were often not enough to overcome less equipped but larger local armies.

Acculturation and miscegenation also diluted native culture and genes. Much like the Roman Empire had assimilated ethnic groups across Europe, Europeans brought their languages, religions, and laws to Native Americans. In some cases, acculturation greatly improved living conditions but it was hardly a choice, as those who refused could face slavery or death. For instance, Spain's *Requerimento* was read to local populations as a justification for the

---

42 "Trans-Atlantic Slave Trade—Estimates", Slave Voyages, retrieved January 2025, https://www.slavevoyages.org/assessment/estimates.

43 See Worse News.

occupation, arguing the Pope had granted them rule over New World lands, and warning of the dangers of rejecting Christ.[44]

There was also acculturation among native groups. European occupation and the introduction of horses disrupted established boundaries and brought together previously isolated peoples. The Mapuche, for instance, expanded from the west of the Andes to Patagonia, absorbing Tehuelche groups and other communities in the process.[45]

The effects of miscegenation are today particularly visible in Brazil. Faced with a vast new territory but a population of little more than a million,[46] Portugal adopted *cunhadismo* as a colonisation tactic. The word *cunhadismo* comes from the Portuguese word *cunhado* (brother-in-law) but it was based on a native tradition. The Tupi, the supra-group inhabiting most of the Brazilian coastline, offered a wife to outsiders to establish a relationship of trust and kinship. The Portuguese settlers—few and almost all male—exploited the native's *cunhadismo* and polygamist practices to grow in number. These mestizos, together with mulattos (mixed African and European descent), are ancestors to nearly 40% of the current Brazilian population.[47]

War, mistreatment, acculturation, and miscegenation materially contributed to the dwindling number of Native Americans, but they were not the main factor. Alongside the soldier's sword and the settler's boot, there was a deadlier force at play: disease.

The timeline for the original settlement of the Americas remains a contentious issue, but recent analyses of ancient genomes suggest all Native American peoples descend from a single group that inhabited Northern Asia in isolation since 18,000 to 22,000 years ago.[48] During the last glacial maximum[49] part of the group crossed Beringia, an ancient land bridge connecting Asia to America, and then slowly spread through the entire continent. Those 20,000 years or so of isolation meant Native Americans had

---

[44] John Michael Francis, *Iberia and the Americas: culture, politics, and history; a multidisciplinary encyclopedia*, Volume 1 (Santa Barbara: ABC-CLIO, 2006), on p. 903.

[45] "Reunión con las Comunidades Mapuche-Tehuelche de Chubut", *Museo de La Plata—Univeersidad Nacional de La Plata*, retrieved January 2025, https://web.archive.org/web/20150416141331/http://www.museo.fcnym.unlp.edu.ar/articulo/2014/9/13/reunion_comunidades.

[46] José Machado, "No Centenário do I Recenseamento Populacional Português", *Revista do Centro de Estudos Demográficos*, 16, 1965, p. 83–104, on p. 87.

[47] Louis Putterman, "World Migration Matrix, 1500–2000", Appendix - Americas 1.1., 2009, https://sites.google.com/brown.edu/louis-putterman/world-migration-matrix-1500-2000.

[48] J. Moreno-Mayar, B. Potter, L. Vinner, L. et al.,"Terminal Pleistocene Alaskan genome reveals first founding population of Native Americans". *Nature* 553, p. 203–207, 2018, on p. 203, https://doi.org/10.1038/nature25173.

[49] Most recent peak in the extent of continental glaciers.

never been exposed to most Old World diseases. In the sixteenth century, those viruses, bacteria, and parasites crossed the Atlantic aboard ships, with devasting effects. In a short period of time, natives were exposed to a wide range of terrifying diseases, including smallpox, measles, bubonic plague, whooping cough, typhus, and malaria.

The exchange was lopsided: from the New World to the Old World, the only major disease to cross the Atlantic was venereal syphilis.[50] Two reasons likely explain this disparity. On the one hand, there was less diversity in the New World. Most life forms, including viruses and other microorganisms, thrive in warmer temperatures, some 40 degrees above and below the equator. Within that latitude range, the New World has 90 degrees of longitude (from California to Brazil), whereas the Old World has 160 degrees (from Dakar to Japan). Less land creates less interactions, less adaptability, and less natural selection. On the other hand, New World peoples had domesticated less animals than their Old World counterparts. Many diseases use these animals as intermediary hosts, so less domestication decreases the chances of human infection and the ensuing buildup of resistances.[51]

Without acquired immunity, the deadliest introduced diseases killed between 25 and 50% of the infected natives, which was not higher than that of first-contact epidemics in the Old World. If diseases were as deadly west of the Atlantic as they had been east of it, why did they nearly wipe out an entire continent? One likely reason is that Native Americans were exposed to multiple new diseases simultaneously, with a devastating combined effect. Another probable contributor was settlement: the native population, rather than having the opportunity to recover over a few generations, was replaced by European settlers and African slaves.[52]

The worst imported disease was smallpox, an illness of many firsts: the first in death toll (with 300 to 500 million fatalities estimated globally[53]), the first to be preventable by a vaccine, and the first (and only one so far) to be eradicated.[54] To be an effective killer, a microorganism needs not only to kill a

---

[50] Kristin Harper et al., "On the origin of the treponematoses: a phylogenetic approach", *Public Library of Science: Neglected Tropical Diseases*, 2(1), 2008, p.1–13, on p.1, https://doi.org/10.1371/journal.pntd.0000148.

[51] Jared Diamond, *Guns, Germs and Steel* (New York: W. W. Norton and Company, 1997), on p. 176–191.

[52] Suzanne Austin Alchon, *A Pest in the Land* [...], on p. 1–5.

[53] Catherine Thèves et al., "The rediscovery of smallpox", *Clinical Microbiology and Infection*, 20(3), 2014, p.210–218, on p.210, https://doi.org/10.1111/1469-0691.12536.

[54] Smallpox is the only human disease declared by the WHO to be eradicated. On animals, Rinderpest, or cattle plague, has also been eradicated.

Table 7.2 Estimated CFR, R_0 and global death toll from pandemics[55]

Disease	CFR	R_0	Death toll
AIDS	80% (untreated)	1.5–4.6	38 million
Bubonic plague	50% - 60%	1.4–1.9	120–260 million
Smallpox	**50%**	**3.5–6.0**	**300–500 million**
Ebola	50%	1.0–1.9	11 thousand (2014–2016)
MERS	40% - 65%	0.3–0.8	800 (2015–2020)
SARS	11%%	0.2–3.9	800 (2002–2003)
Spanish flu (H1N1)	2% - 3%	1.4–3.8	17–50 million (1918–1919)
COVID-19	0.8% (2024)	3 (2024)	7–35 million (2019–2024)
Measles	0.3% - 5%	12–18	2.6 million per year (by 1963)

large portion of its hosts, but also to be highly infectious. These two characteristics are denoted by two numbers: the CFR (Case Fatality Rate, or the proportion of infected patients who die from the disease) and the R_0 (basic reproduction number, or the number of individuals infected by each patient). Glancing at these two parameters across pandemics paints a stark picture of the viciousness of smallpox. Compared to other diseases, it is as deadly as bubonic plague but more contagious, and less contiguous than measles but far deadlier (Table 7.2).

There are no certainties about the origins of smallpox but it is undoubtedly an old disease, with references to its symptoms appearing in Indian texts from c1500 BC and Chinese manuscripts from c1100 BC. The wreckage it left throughout the ages has been amply documented: 25% of an army dead in Athens, 430 BC; 25% to 33% death rate in the Roman Empire, 165–180 AD; 30% to 90% decline in male taxpayers in Egypt, 165–180 AD; 44% of the adult population dead in Izumi, Japan, 735–737 AD.[56]

Smallpox's variola virus is mostly transmitted through direct person-to-person contact during the up to four-week period the patient is infectious. Indirect transmission via clothing and personal items is also possible, but

---

[55] Adapted from: Silvio Daniel Pitlik, "COVID-19 Compared to Other Pandemic Diseases", *Rambam Maimonides Medical Journal*, 11(3), 2020, https://doi.org/10.5041/RMMJ.10418; Jean-Pierre Unger, "Comparison of COVID-19 Health Risks With Other Viral Occupational Hazards", *International Journal of Health Services*, 51(1), 2021, p. 47–49, https://doi.org/10.1177/0020731420946590; Jieliang Chen, "Pathogenicity and transmissibility of 2019-nCoV-A quick overview and comparison with other emerging viruses", *Microbes and Infection*, 22(2), 2020, p. 69–71, https://doi.org/10.1016/j.micinf.2020.01.004; Grace Patterson et al., "Societal Impacts of Pandemics: Comparing COVID-19 With History to Focus Our Response", *Frontiers in Public Health*, 9, 2021, https://doi.org/10.3389/fpubh.2021.630449; Patrick Berche, "History of measles", *La Presse Médicale*, 51(3), 2022, https://doi.org/10.1016/j.lpm.2022.104149; WHO, "Data", 2024, https://data.who.int/; The Economist, "Tracking covid-19 excess deaths across countries", July 2024, https://www.economist.com/graphic-detail/coronavirus-excess-deaths-tracker.

[56] Suzanne Austin Alchon, *A Pest in the Land* […], on p. 21.

in tropical climates the virus does not survive for more than four weeks.[57] Magellan's fleet took about six weeks to cross the Atlantic, so it could not have brought smallpox to Rio de Janeiro. However, by then the virus had already managed to leap over the ocean. After Columbus' first voyages and with increasing awareness of trade winds, ships started to routinely cross the Atlantic in under four weeks, so it was a matter of time until smallpox reared its ugly head in the New World. The first reported epidemics were in the island of Hispaniola, in 1507 and 1517, from where it spread to Mexico in 1520, causing countless deaths among the Aztecs and other native groups. Trade routes and additional infected European ships eventually distributed the disease to the entire continent.[58]

Smallpox is not a subtle disease. After a period of incubation of 10 to 15 days, the patient develops incapacitating fever, vomiting and pain. Three to five days after, the first macules appear, progressively evolving to papules, vesicles, and then to pustules. Often in the hundreds, these can fully cover the face and forearms of the patient. Death is caused by cytopathic effects (widespread cell breakdown caused by the viral infection). Those that survive gain long lasting immunity[59] but are left with pockmarks and, in some cases, blindness.[60]

The Chinese had discovered that sniffing dried scabs from a mild case of smallpox could offer immunisation against the disease. The Arabs did something similar, placing the dried scabs in contact with a scratch. Variolation, as these methods came to be known, was practiced in Asia since at least the sixteenth century, but only definitely made its way to Europe and the New World in the eighteenth century,[61] much too late for the Native American population.

Variolation was also not risk-free. Patients became infectious and had to be kept in isolation, not all acquired life-long immunity, and about 2% of them died from the procedure. As the end of the eighteenth century neared, Edward Jenner, a British physician, found a better alternative. Jenner was a countryside doctor, and he noticed that some of his farmer patients refused

---

[57] Abbas Behbehani, "The smallpox story: life and death of an old disease", *Microbiological Reviews*, 47(4), 1983, p. 455–509, https://doi.org/10.1128/mr.47.4.455-509.1983, on p. 507.

[58] Dauril Alden and Joseph Miller, "Out of Africa: The Slave Trade and the Transmission of Smallpox to Brazil, 1560–1831", *The Journal of Interdisciplinary History*, 18(2), 1987, p. 195–224, on p. 198, 199.

[59] Erika Hammarlund et al., "Antiviral Immunity following Smallpox Virus Infection: a Case–Control Study", Journal of Virology, 84(24), 2010, p. 12,754–12,760, on p.12754, doi.org/10.1128/jvi.01763-10, https://doi.org/10.1128/jvi.01763-10.

[60] Abbas Behbehani, "The smallpox story: life and death of an old disease" […], on p. 483.

[61] Arthur Boylston, "The origins of inoculation", *Journal of the Royal Society of Medicine*, 105(7), 2012, p.309–313, on p.309, https://doi.org/10.1258/jrsm.2012.12k0

variolation, as they held the belief that previous cowpox infections made them immune to smallpox. Jenner, who had himself nearly died from variolation at the age of eight, was intrigued. He pursued the idea and found cowpox inoculations could indeed protect from smallpox, and so the first vaccine was created. The name Jenner chose for his creation betrays its roots, as *vacca* is Latin for cow.[62]

However, progress in defeating the disease was slow. The vaccine proved challenging to produce and distribute in large numbers, and an estimated 300 million people still died from smallpox in the twentieth century, nearly three times as much as the combined casualties of the first and second world wars. In 1958, the World Health Organization proposed an ambitious plan for the global eradication of the variola virus, which required vaccinating about 80% of the population where the disease was still endemic (most of Asia, Africa, and South America). The initial results of the programme were disappointing, but winds started to shift in the next decade, with additional funding and a concerted effort to mass produce 200 million vaccines per year. Vaccination rates were reduced from 80 to 50% of the population at risk, but epidemiological surveillance procedures were improved. The new plan leveraged the fact that smallpox spreads slowly so new outbreaks could be, if detected early, contained with targeted vaccination. The strategy worked and, in 1975, the last case of smallpox was reported in Bangladesh.[63] The patient, a 3-year-old Bangladeshi girl named Rahima Banu, survived the ordeal.[64]

Regrettably, that would not be quite the end of it. In 1978, Janet Parker, a medical photographer in the University of Birmingham died from smallpox, when the virus escaped from a nearby ill-equipped research laboratory.[65] The incident prompted the destruction of all known samples of variola virus, except for those kept in the CDC (Atlanta, US) and the Vector Institute (Koltsovo, Russia), two BSL-4 facilities (the highest standard for biosafety).[66]

The World Health Organization and several individual countries still keep stockpiles of smallpox vaccines for the event of a new outbreak. The possibility of a lab leak—either accidental or deliberate—has raised concerns. The destruction of these lingering samples is routinely discussed, but even that

---

[62] Nancy Männikkö, "Etymologia: Variola and Vaccination", *Emerging Infectious Diseases*. 2011. 17(4), p.680, https://doi.org/10.3201/eid1704.et1704.

[63] An ensuing case of variola minor (a milder strain of the virus) was reported in Somalia, in 1977 (Donald Henderson, "The eradication of smallpox–an overview of the past, present, and future", *Vaccine*, 29(4), 2011, p. D7–D9, on p. D9).

[64] Abbas Behbehani, "The smallpox story: life and death of an old disease" […], on p. 496.

[65] *Report of the investigation into the cause of the 1978 Birmingham smallpox occurrence* (London: Her Majesty's Stationery Office, 1980), on p. 38.

[66] CDC, "History of Smallpox", May 2024, https://www.cdc.gov/smallpox/history/history.html.

would not eliminate the possibility of malicious intent, as the variola virus can arguably be synthetized from the publicly available full genome.[67]

It would certainly not be the first time smallpox was used as a biological weapon. The first references date back to the fourteenth century, when Tatar forces catapulted disease riddled corpses into encircled towns.[68] The New World also witnessed its share of this type of warfare. In the nineteenth century, a Brazilian native tribe of Timbira was lured to Caxias, a smallpox-infested settlement, to clear their lands for cattle ranching.[69] Before, in the eighteenth century, British forces found themselves besieged in a fort in present-day Pittsburgh, surrounded by Lenape natives. Two blankets and a handkerchief were infected with the variola virus and presented as a gift to the Lenape emissaries. The idea was proposed by General Jeffrey Amherst to the fort's commander in the following terms:

> Could It not be contrived to send the smallpox among the disaffected tribes of indians? We must on this occasion use every stratagem in our power to reduce them.[70]

---

[67] Ryan Noyce et al., "Construction of an infectious horsepox virus vaccine from chemically synthesized DNA fragments", *PloS ONE, 13(1), 2018, p. 1–16, on p. 1,* https://doi.org/10.1371/journal.pone.0188453.

[68] Edward Janoff and Ruth Lynfield, "Smallpox: remembrance of things past, or the coming plague?", *The Journal of Laboratory and Clinical Medicine,* 142(4), p.211–215, on p.211, https://doi.org/10.1016/S0022-2143(03)00152-5.

[69] Darcy Ribeiro, *O Povo Brasileiro: a formação e o sentido do Brasil* […], on p. 42.

[70] American Society for Microbiology, "Investigating the Smallpox Blanket Controversy", 2023, https://asm.org/articles/2023/november/investigating-the-smallpox-blanket-controversy.

8

A Bad Ending to a Great Start

From Santa Cruz (October 18th, 1520) to the Finding of the Magellan Strait (Late November, 1520)

The bleak winter was behind them, but time was not on their side. While the two months spent in Santa Cruz had allowed the fleet to stock up on fresh water and smoked meat, the supplies of biscuit and wine—less perishable and therefore the basis of a sailor's diet[1]—had been dwindling since the ships had left Spain, more than a year ago.

After three days of sailing down the coast, they found what looked like a bay (Fig. 8.1). It was large and deep, so Magellan ordered Carvalho, the *Concepción*'s pilot, to take a crew ashore and to the top of a small hill on the northern tip of the bay. They called the tip *Punta Vírgenes*, in the honour of the feast of the Eleven Thousand Virgins of St. Ursula, celebrated that day.[2] Today it is known instead as *Punta Dungeness*, and marks the border between Argentina and Chile. After returning aboard, Carvalho reported the bay looked closed to the west. Magellan, unconvinced, sent the *San Antonio* and the *Concepción* to take a closer look, under strict instructions to spend no more than five days investigating, given the shortage of rations.[3]

Ensuing Spanish charts for this region would advise sailors that there are no good seasons here, as even spring and summer can unexpectedly churn violent squalls. The fleet experienced this first-hand when a sudden gale arose from the northeast. The *Trinidad* and the *Victoria* anchors started to drag, so Magellan directed the two ships out of the bay. The storm quickly subsided,

---

[1] See What Did 16th Century Sailors Eat?

[2] Antonio Pigafetta, *The first voyage around the world 1519–1522*, edited by Theodore Cachey Jr. (Toronto: University of Toronto Press, 2007), on p. 18.

[3] Visconde de Lagôa, *Fernão de Magalhães: A sua vida e a sua viagem* (Lisboa, Seara Nova, 1938), on Volume II, p. 223.

R. Gaspar, *The Revolution of Magellan*, https://doi.org/10.1007/978-3-032-10797-8_8

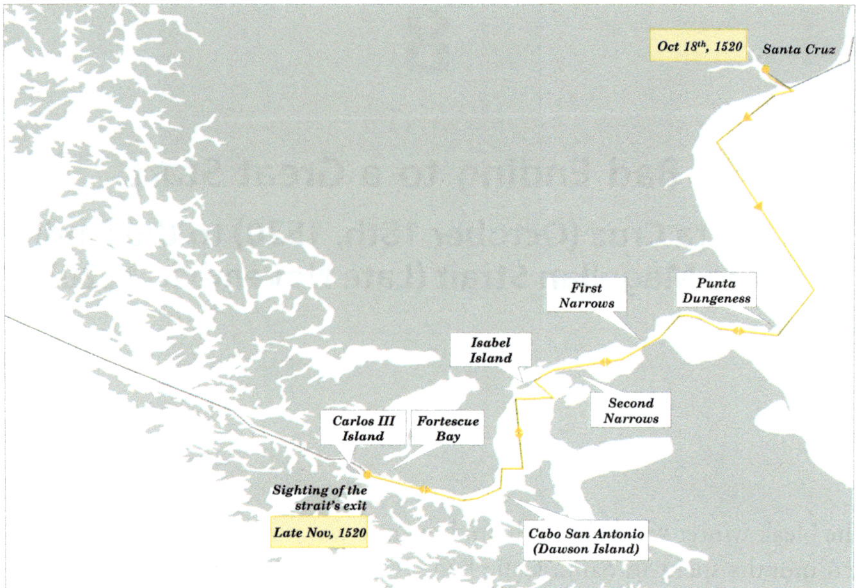

Fig. 8.1 Fleet route[4]

but not before they lost sight of the *San Antonio* and the *Concepción*. They were already deep into the bay when the wind had picked up, and Magellan feared they had no room to outmanoeuvre it before being thrown onto the shore.[5]

After an excruciating wait, the two ships appeared right at the end of the allotted five days, with guns blazing and masts filled with banners. It was the first outburst of happiness since the good days spent in Rio de Janeiro, and it marked a crucial point in the expedition: Estevão Gomes, the *San Antonio's* pilot, was convinced they had found a passage to the west. When the storm hit, they were indeed too deep into the bay to battle their way out against the northeaster wind. Unable to anchor, they found themselves hopelessly dragged towards the bay's southwest sand banks. While they were desperately tacking northwest to avoid grounding the ship, they spotted a canal to the west, which they promptly followed. After about 4 km, the canal— today called the First Narrows—opened to a large bay, where the *San Antonio* weathered the rest of the storm. When it subsided, the ship further explored westward, crossing a second canal—the Second Narrows—into a wide gulf

---

[4] Fleet route adapted from Tomás Mazón Serrano ("Mapas", Ruta Elcano, June 19, 2024, https://rutaelcano.com/mapas/); map adapted from Natural Earth (1:10 m large scale land data, version 5.1.1, https://www.naturalearthdata.com/downloads/).

[5] Tim Joyner, *Magellan* (Maine: International Marine, 1992), on p. 153.

which continued south between the increasingly higher banks. They had no time to continue investigating, but the deep and salty waters, plus the balanced ebb and flood tidal currents, convinced Gomes that they were in an oceanic passage.[6]

The fleet retraced the *San Antonio's* route and, after crossing the Second Narrows, found shelter between Isabel Island and the shore. Magellan convened a council of captains and pilots to decide whether they should continue forward or return to Spain.[7] Was Magellan finally departing from his autocratic leadership style? Probably not. Knowing a passage had been found, the captain general could now afford to show—or feign—some malleability to distance himself from what had happened at Port San Julián. Perhaps sensing insincerity, all officers were in favour of continuing, with one exception. Gomes—with his confidence boosted after discovering the passage—argued it was risky to continue, as the journey ahead was long and the supplies short. It would be wiser, he insisted, to return to Spain and organise a second expedition to continue to explore the newly discovered passage.[8]

It seemed indeed like the most rational thing to do, as it would safeguard the crew and the king's ships. It would do little for Magellan, though. The passage, while crucial, was just a fraction of what he had committed to do. Faced with but a timid success and a barrage of controversial events, it was unlikely the monarch would allow Magellan to lead a follow-up expedition. After hearing Gomes' objections, Magellan is reported to have replied that *"even if they had to eat the leather on the ships' yards, he would still go on, to discover what he had promised to the king"*.[9] If Magellan indeed said it, he was being remarkably prescient.

Reinforced by the apparent near concord, Magellan ordered the fleet to raise anchors and sail down the large channel. They made steadfast progress and happily noted the barren landscapes of earlier giving way to greener and more hospitable surroundings. It was November 1st, so he named the passage All Saints Channel,[10] but a few years later it would already be known as the Strait of Magellan.[11] During the day the crew could see puffs of smoke rising

---

[6] Idem.

[7] José Toribio Medina, *El descubrimiento del Océano Pacífico: Vasco Núñez de Balboa, Fernando de Magallanes y sus compañeros* (Santiago de Chile: Imprenta Universitaria, 1920), on p. CCXLVI.

[8] Mateo Martinic, *Historia del Estrecho de Magallanes* (Santiago de Chile: Editorial Andres Bello, 1977), on p. 44.

[9] Idem.

[10] Mateo Martinic, *Historia del Estrecho de Magallanes* [...], on p. 43.

[11] Joaquim Gaspar and Šima Krtalić, *The Cartography of Magellan* (Lisbon: Tradisom, 2023), on p. 168–180.

above the shoreline and, when the evening fell, campfires also became visible here and there. So, they reportedly baptised these coasts as *Tierra de Los Fuegos* (Land of The Bonfires), later adapted to its current name of *Tierra del Fuego* (Land of The Fire).[12]

They were presently at a fork in the strait and could see three passages continuing into the distance. Magellan's flagship, together with the *Victoria* and the *Concepción*, took the most promising southwestern passage, while the *San Antonio* was given orders to explore the other two channels. After three days, the fleet would regroup at what is today called Cape San Antonio, in Dawson Island—likely the furthest tip of the southwestern passage which is visible from the fork where the fleet split up.[13]

Once Magellan reached Cape San Antonio, he could see the strait turning sharply to the right and continuing into the northwest. Could this be finally the passage that would lead them to the vast *Mar del Sur* ("South Sea") Balboa had seen in 1513, when he crossed the Isthmus of Panama[14]? Not wanting to lose time, Magellan sent the *Concepción* to fetch the *San Antonio* and sped northwest with the *Victoria*. After sailing for about 50 km, the *Trinidad* spotted what they called the "River of Sardines" (modern-day Fortescue Bay)[15] on the northern shore of the passage. With its sheltered port, fresh water supplies, and indeed plenty of sardines, it offered a safe harbour for the fleet to reconvene and prepare for the long oceanic crossing ahead. Twenty kilometres ahead of Fortescue Bay, the Carlos III Island blocks the view of the rest of the passage, so Magellan sent a longboat to check what lied beyond it. The small crew returned three days after, with news worthy of *albricias* (a monetary reward customarily given by the Spanish Crown to the bearers of exceptional news): they had circled past the island and—perhaps after climbing a hill—seen the great South Sea![16]

Alas, not everything was good news, as there was still no sign of the *Concepción* and the *San Antonio*. After nervously waiting one more day, Magellan ordered the *Trinidad* and the *Victoria* to go back east seeking the two missing ships. They found the *Concepción* returning, bearing worrisome news: the *San Antonio* was nowhere to be found. The three ships spent the next six days sailing up and down the strait, in pursuit of the missing companion. The *Trinidad* and the *Concepción* combed both sides of Dawson Island, while the *Victoria* travelled all the way back to *Punta Dungeness*, at the entrance to the

---

¹² Mateo Martinic, *Historia del Estrecho de Magallanes* [...], on p. 43.

¹³ Tim Joyner, *Magellan* [...], on p. 157.

¹⁴ See How Did Magellan Convince the Spanish King?

¹⁵ Mateo Martinic, *Historia del Estrecho de Magallanes* [...], on p. 45.

¹⁶ José Toribio Medina, *El descubrimiento del Océano Pacífico* [...], on p. CCXL, CCXLI.

Atlantic. Finding no evidence of the ship, they left messages for the missing crew in prominent places, each one marked with a large stone cairn and a wooden cross.[17]

The disappearance of the *San Antonio* was disconcerting. There had been no major storms since the fleet split up, and they had found no traces of a wreck nor marooned sailors. Could the fleet's largest ship have sunk in the strait's deep waters without a trace? Confused, Magellan asked San Martín, the fleet's astronomer *and* astrologer, to cast a horoscope. After consulting the stars, San Martín grievously announced the *San Antonio* had deserted back to Spain.[18]

San Martín's horoscope was dead on. The *San Antonio* had indeed deserted, with the uncooperative captain Mesquita in chains. Assuredly San Martín did not read this in the stars, but rather deducted it from the persisting anxiety and fear he could sense in the crew. Probably only Magellan—lost in the excitement of having discovered the strait—did not realise that most of his crew would probably follow the *San Antonio*, if given the chance.

This time, the uprising had not been stirred by Spaniards but rather by Gomes, the Portuguese pilot. Gomes had accompanied Magellan when he moved to Spain, but they never seemed to fully trust each other. Gomes had even presented his own plan for a western route, but the Spanish Crown preferred Magellan's.[19] Still, nobody doubted his navigation skills and he secured the position of pilot-major on Magellan's roster. At Rio de Janeiro, his pride was most likely hurt when Magellan gave the command of the *San Antonio* to Mesquita, a less experienced yet loyal relative, and demoted Gomes from his position as pilot-major of the flagship to pilot at the *San Antonio*, serving under Mesquita.

Alone at the *rendezvous* point, surrounded by unfamiliar snow-capped mountains and with shrinking provisions, it did not take much for Gomes to convince the crew to seize Mesquita and turn back to Spain before Magellan's determination to press onward killed them all. They sailed directly back to Spain without even stopping at Port San Julián to look for Cartagena and Calmette and, by May 6th, six months after deserting, the *San Antonio* was back in Seville. The captive Patagonian travelling in the ship, likely not the first in line for the distribution of the rationed provisions, was not among the 55 men who made it back.[20]

---

[17] Idem, on p. CCXLV, CCXLVI.

[18] Idem.

[19] Tim Joyner, *Magellan* [...], on p. 82.

[20] José Toribio Medina, *El descubrimiento del Océano Pacífico* [...], on p. CCXLVIII–CCLII.

Knowing that the Spanish authorities would not take desertion lightly, the crew signed a statement accusing Magellan of mistreatment and murder at Port San Julián, forcing Mesquita to sign it too.[21] The statement would also hopefully deflect Bishop Fonseca's rage to Magellan, minimising the amount of grief they would receive from having left Fonseca's relative Cartagena stranded in Patagonia. In any case, this decision may have not been so callous as it appears, as the *San Antonio* could have been merely following trade winds, which would lead them northeast rather than north. A c1521-1524 chart, made by the Reinel using reconnaissance data brought by the ship, contains the first know representation of the Falkland Islands[22] (which are northeast of the strait), suggesting this was indeed the case.

The plan worked. Although the *San Antonio's* officers were imprisoned the moment they set foot in Seville, all of them were later released except Mesquita. Despite claiming he had signed the statement under duress and presenting records for the court martial he presided at Port San Julián—which he had kept hidden while shackled aboard the *San Antonio*—his version of events was deemed unconvincing. As the deserters had hoped, Bishop Fonseca focused his rage on Magellan rather than on them, and even tried to take into custody the captain general's wife and child.[23]

8.1   How Do We Know What Happened in the Expedition?

The short answer

If the statements from the deserters were the only surviving testaments, today we would surely have a very different view of this voyage. Fortunately, this was one of the best documented 16th century expeditions, with more than 10 surviving accounts from eyewitnesses. There are also dozens of secondary contemporary records, written by those who talked to eyewitnesses or had access to their journals, a score of bureaucratic records, and the studies of the many historians who have pored over these resources in the last centuries. Of course, with a wealth of information comes the challenge of resolving conflicting information. Often such divergences can be settled with a reasonable degree of confidence, but for some episodes of Magellan's journey we will perhaps never know what truly happened.

---

[21] Tim Joyner, *Magellan* [...], on p. 158.

[22] Joaquim Gaspar and Šima Krtalić, *The Cartography of Magellan* (Lisbon: Tradisom, 2023), on p. 150–163.

[23] José Toribio Medina, *El descubrimiento del Océano Pacífico* [...], on p. CCLIV.

Had the Spanish authorities been less predisposed to condemn Magellan and his relative Mesquita, they would probably have given more importance to inconsistencies in the statements provided by the *San Antonio* crew. For instance, they claimed Magellan and Mesquita had punished San Martín with *tratos de cuerda* after the Port San Julián rebellion.[24] This horrible form of punishment, quite popular in the sixteenth century, involved tying the subject's arms behind their back with a rope, which was then used to hoist them. Once dangling in the air, the victim was suddenly dropped, inevitably dislocating their shoulders and often leaving them disabled for life. It would be odd for Magellan to risk crippling the only member of his crew who could determine longitude, a skill needed to prove the Moluccas were on the Spanish hemisphere. A couple of weeks after the alleged torture, San Martín measured the longitude of Port San Julián,[25] something he would find very hard to do if recovering from dislocated collarbones or worse. The deserters also stated that San Martín was subject to this punishment not only for having rebelled against Magellan—which he did—but also for keeping detailed records of their route.[26] This would be an odd thing for Magellan to be mad about, as it was part of the king's instructions.[27] In fact, the ship's inventory included 24 blank pieces of vellum to chart new territories[28] and other pilots also kept detailed records of their route.

In 1522, testimonies from the *Victoria* survivors[29] would again bring up the issue of contradicting accounts. While some of the professional mariners seemed unprejudiced, it was clear most survivors were either staunchly pro-Magellan or unwaveringly pro-Spanish officers, and that their testimonies were distorted accordingly. Unwilling to tarnish the expedition with a drawn-out inquiry, the crown chose not to look too closely. Magellan and Mesquita were cleared from the accusations of acts of cruelty at Port San Julián, and no further charges were brought against the mutineers of Port San Julián and the deserters of the *San Antonio*.[30]

---

[24] "Carta de los oficiales de la Casa de la Contratación de las Indias al emperador Carlos V" (Seville: Archivo General de Indias, ES.41091.AGI//PATRONATO,34,R.14, 1521), https://pares.mcu.es/ParesBusquedas20/catalogo/description/122223.

[25] See Were San Martín's Longitude Measurements Accurate?

[26] "Carta de los oficiales de la Casa de la Contratación de las Indias al emperador Carlos V" [...].

[27] Cristóbal Bernal, *Crónicas de la Primera Vuelta al Mundo, según sus Protagonistas*, 2016, on p. 142–199.

[28] Martin Fernandez de Navarrate, *Coleccion de los viajes y descubrimientos que hicieron por mar los españoles desde fines del siglo XV*, Tomo 4 (Madrid: Imprenta Nacional, 1837), on p. 180.

[29] See The Lucky Seven Percent Get Home.

[30] See How Did Magellan's Expedition Change the World?

Fortunately, over the centuries many have examined the expedition closely. While some aspects will undoubtedly remain shrouded in mystery forever, there is much that can be pieced together from the countless surviving records, albeit always keeping an eye out for the inconsistencies and biases of their authors.

From all the first-hand accounts, Pigafetta's is the most detailed one, as he kept a diary of the expedition which he later turned into a book. Sadly, his original diary and book manuscript vanished. Four early copies of the book did survive, but these copies are translations which may have flattened nuances in the process.[31] Furthermore, while converting his diary into a book Pigafetta analysed additional documents and discussed the expedition with others,[32] which may have led to changes in his original accounts of events. Pigafetta's reporting should also be taken with a pinch of salt, as the Italian gentleman had a taste for exaggeration and would happily write down tall tales as facts. Pigafetta contributed to the myth of Patagonian giants which endured in Europe for more than 250 years,[33] and later in the expedition he would write about Southeastern Asians with giant ears used as blankets.[34] But perhaps his less obvious hyperboles are the most dangerous ones. Pigafetta developed a close relationship with Magellan throughout the expedition, one that seemingly prevented him from being unbiased when describing conflicts involving the captain general. For Pigafetta, Magellan was a courageous and selfless hero who could do no wrong, while his enemies were jealous fools.

Francisco Albo, a professional Greek mariner known both for his experience and for not getting involved in political disputes, compiled the expedition's most detailed course logbook, used as the main source to retrace the fleet's wanderings. His logbook[35] remained undiscovered in Spanish archives until 1788 (well after the end of Portugal and Spain's disputes over the Moluccas), reducing the likelihood of tampered information (conversely,

---

[31] Several authors have published annotated transliterations and translations of these manuscripts, see for instance: Antonio Pigafetta, *The first voyage around the world 1519–1522*, edited by Theodore Cachey Jr. (Toronto: University of Toronto Press, 2007).

[32] Teresa Carvalho, "Documenting the tropical natural world in the account of Antonio Pigafetta", *Magallánica: Revista de Historia Moderna*, 7(13), 2020, p. 288–314, on p. 291.

[33] See Winter in Patagonia.

[34] Antonio Pigafetta, *The first voyage around the world 1519–1522* [...], on p. 116, 117.

[35] Transliterated and annotated by Cristóbal Bernal (*Crónicas de la Primera Vuelta al Mundo* [...], on p. 381–433), also available in English (H.E.J. Stanley, *The first voyage round the world, by Magellan. Translated from the accounts of Pigafetta, and other contemporary writers*, London, Hakluyt Society, 1874, on p. 211–236).

the longitude readings Pigafetta included in his book questionably placed the entire Southeast Asia in Spain's hemisphere[36]).

The remaining first-hand accounts of the expedition are less comprehensive than Pigafetta and Albo's, but nevertheless useful for filling in gaps and double-checking facts. The *Genoese Pilot*, likely León Pancaldo,[37] would assume the piloting duties for the *Trinidad* after it left the Moluccas.[38] Ginés de Mafra, an able sailor, complemented Pancaldo's account of that trip. Mafra also included information seemingly taken from San Martín's navigational papers, which he had in his possession.[39] Vasquito Galego, an apprentice sailor and son of Vasco Galego, the *Victoria's* first pilot, is the likely author of an anonymous account dictated to a scribe several decades after the expedition.[40] Martín Mendez, the *Victoria's* clerk, wrote a description of the pacts made with rulers in the Philippines and Moluccas.[41] The list goes on, but these accounts are the best-known sources of first-hand information about the expedition. There is of course always the hope of finding new writings which remain hidden by time. Martín Mendez's papers, for instance, were only discovered in the twentieth century, after staying buried in Spanish archives for four centuries.

On top of these first-hand accounts there are also several secondary accounts, written by those who had a chance to talk to members of the expedition. The rather large group includes Álvaro da Costa and Sebastião Álvares, respectively the Portuguese ambassador and consul that attempted to dissuade Magellan from the enterprise[42]; Bartolomé de las Casas, a missionary who was present in Magellan's audience with the Spanish king and later had an opportunity to speak with the navigator[43]; Pedro Mexia, a court chronicler that witnessed the departure of the ships from Seville and the return of

---

[36] Rolando Laguarda Trías, "Las longitudes geográficas de la membranza de Magallanes y del primer viaje de circunnavegación", *Actas do II Colóquio Luso-Espanhol de História Ultramarina*, 1975, p. 135–174, on p. 171.

[37] José Manuel Garcia, "Documentos existentes em Portugal sobre Fernão de Magalhães e as suas viagens", *Abriu*, 8, 2019, p. 15–33, on p. 20.

[38] H.E.J. Stanley, *The first voyage round the world* [...], on p. 1–20.

[39] Antonio Blazqez and Delgado Aguilera, *Libro que trata del descubrimiento y principio del Estrecho que se llama de Magallanes, por Gines de Mafra* (Madrid: Real Sociedad Geográfica, 1921).

[40] H.E.J. Stanley, *The first voyage round the world* [...], on p. 30–32.

[41] Mauricio Obregón, *La primera vuelta al mundo: Magallanes, Elcano y el Libro perdido de la nao Victoria* (Bogotá: Plaza & Janés, 1984).

[42] See, for instance, Sebastião Álvares' letter describing his encounter with Magellan: "Carta de Sebastião Álvares sobre o seu encontro com Fernão de Magalhães" (Lisbon: Arquivo Nacional da Torre do Tombo, Corpo cronológico 1-13-20, 1519).

[43] Bartolomé de Las Casas, *Historia de Las Indias*, Tomo IV, ed. Marqués de La Fuensanta Del Valle and D. José Sancho Rayon (Madrid: Imprenta de Miguel Ginesta, 1876).

the *Victoria*, three years later[44]; López Recalde and Sancho de Matienzo, the *Casa's* officials who interviewed the *San Antonio* deserters[45]; Peter Martyr of Angleria,[46] Maximilianus Transilvanus,[47] and other court officials that interviewed survivors from the *Victoria*; and António de Brito, the Portuguese commander who seized the *Trinidad* in the Moluccas.[48]

Then there are the Portuguese and Spanish chroniclers with access to expedition documents which have since been lost. Given the two kingdom's dispute over the Moluccas, it is perhaps no surprise these second-hand accounts were often biased. On the Portuguese faction, João de Barros[49] and Fernão Lopes de Castanheda[50] had ample access to relevant expedition records, including the correspondence between Magellan and Serrão, the papers confiscated from the *Trinidad*, and the longitude and navigational annotations of San Martín. None of the two chronicles make much of an effort to disguise their disgust for Magellan, rooted in his decision to take Portuguese charts and serve the Spanish Crown. Damião de Gois, a Portuguese chronicler responsible for documenting King Manuel I's reign and who also had access to records from the expedition, seemed less prejudiced.[51] Other Portuguese contemporary writers that discussed Magellan and his expedition include Gaspar Correa[52] and António Galvão.[53]

Of the Spanish chroniclers, Gonçalo Fernández de Oviedo y Valdeés discusses the expedition but does so based solely on a report written by Elcano.[54] Towards the end of the sixteenth century, Antonio Herrera, an

---

[44] Pedro Mexia, *Dialogos,* Coloquilo del Sol (Madrid: Imprenta de Francisco Xavier Garcia, 1767), on p. 138–140.

[45] Manuscript: "Carta de los oficiales de la Casa de la Contratación de las Indias al emperador Carlos V" […]; Transliteration: Cristóbal Bernal, *Crónicas de la Primera Vuelta al Mundo* […], on p. 329–380.

[46] Francis MacNutt, *De Orbe Novo: The Eight Decades of Peter Martyr D'Anghera* (New York, G.P. Putnam's Sons, 1912).

[47] H.E.J. Stanley, *The first voyage round the world* […], on p. 179–210.

[48] "Carta de António de Brito com notícias sobre os castelhanos que estavam nas Molucas, incorporada na carta de Jorge de Albuquerque escrita em Malaca a 28 de agosto de 1522" (Lisbon: Arquivo Nacional da Torre do Tombo, Corpo cronológico 111-15-81, 1522).

[49] João de Barros, *Década terceira da Ásia de João de Barros* (Lisboa: Jorge Rodriguez, 1628).

[50] Fernão Lopes de Castanheda, *História do descobrimento & conquista da Índia pelos portugueses* (Coimbra: João Barreira and João Álvares, 1554).

[51] Damião de Gois, *Chronica do Felicissimo Rei Dom Emanuel composta per Damiam de Goes* (Lisbon: Francisco Correa, 1566–1567).

[52] Gaspar Correa, *Lendas da Índia* (Lisbon: Typografia da Academia Real das Ciências, 1860).

[53] António Galvão, *Tratado que compôs o nobre & notauel capitão Antonio Galuão, dos diuersos & desuayrados caminhos, por onde nos tempos passados a pimenta & especearia veyo da India ás nossas partes* (Lisbon: Ioam da Barreira, 1563).

[54] Included in José Toribio Medina, *El descubrimiento del Océano Pacífico: Vasco Núñez de Balboa, Fernando de Magallanes y sus compañeros* (Santiago de Chile: Imprenta Universitaria, 1920).

official court chronicler, also had a chance to read many of the fleet's records, including San Martín's papers before they vanished.[55] Other relevant contemporary writers include Bartolomé Argensola[56] and Francisco López de Gómara.[57]

While first and second-hand statements are inevitably prone to biases and interpretations, bureaucratic archives are often truer to form. The Spanish Court and the *Casa* stored a nearly bottomless score of contemporary records and correspondence, including Charles V approval of the expedition, his lengthy set of rules for Magellan, the original crew rosters and lists of the deceased, the expedition's tally of costs and items taken aboard, and the final profits from the sale of cloves.[58] Portugal too kept records concerning Magellan and his expedition, including his formal requests to King Manuel I, snippets of his military career in the Indies, and official correspondence of the endeavours to derail the expedition.[59]

Over time, many historians have studied Magellan's expedition in depth, offering compilations of the original sources and a view on how to untangle contradictions. Notable mentions include Spain's Martín Fernández de Navarrete in 1837[60]; Britain's Henry Stanley in 1874,[61] Chile's José Toribio Medina in 1888[62] and 1920,[63] Britain's Francis Guillemard in 1890[64]; Belgium's Jean Denuncé in 1911[65]; Portugal's João Júdice (Visconde de

---

[55] Antonio Herrera, *Historia general de los hechos de los castellanos en las islas i Tierra Firme del Mar Oceano* (Madrid: En la emplenta Real, 1601).

[56] Bartolomé Argensola, *Conquista de las islas de Maluco al rey Felipe III* (Madrid: Alonso Martín, 1609).

[57] Francisco López de Gómara, *Historia General de las Indias* (Zaragoza: Agustín Millán, 1552).

[58] The main documents have been transliterated by Cristóbal Bernal (*Crónicas de la Primera Vuelta al Mundo* [...]), and the original manuscripts are stored in Seville's Archivo General de Indias (accessible online at PARES, https://pares.cultura.gob.es/inicio.html).

[59] José Manuel Garcia has compiled a listo of the most relevant documents ("Documentos existentes em Portugal sobre Fernão de Magalhães e as suas viagens", Abriu, 8, 2019, p. 15–33), most of which are stored in Lisbon's Arquivo Nacional da Torre do Tombo (accessible online at https://antt.dglab. gov.pt/pesquisar-na-torre-do-tombo/).

[60] Martín Fernández de Navarrete, *Coleccion de los viajes y descubrimientos que hicieron por mar los españoles desde fines del siglo XV* (Madrid: Imprenta Nacional, 1837).

[61] H.E.J. Stanley, *The first voyage round the world, by Magellan. Translated from the accounts of Pigafetta, and other contemporary writers* (London: Hakluyt Society, 1874).

[62] José Toribio Medina, *Colección de documentos inéditos para la historia de Chile: desde el viaje de Magallanes hasta la batalla de Maipo: 1518–1818* (Santiago de Chile: Imprenta Ercilla, 1888).

[63] José Toribio Medina, *El descubrimiento del Océano Pacífico: Vasco Núñez de Balboa, Fernando de Magallanes y sus compañeros* (Santiago de Chile: Imprenta Universitaria, 1920).

[64] Francis Guillemard, *The life of Ferdinand Magellan and the first circumnavigation of the globe: 1480–1521* (New York: Dodd Mead, 1890).

[65] Jean Denucé, *Magellan: la question des Moluques et la première circumnavigation du globe* (Brussels: Hayez, 1911).

Lagôa) in 1938[66]; United States' Charles Nowell in 1962[67]; United States' Samuel Eliot Morison in 1974[68]; and United States' Tim Joyner in 1992.[69] Joyner, an oceanographer and World War II veteran who travelled through the Strait of Magellan, continues to have one of the better grounded and enjoyable accounts of the expedition, despite more recent attempts.

[66] Visconde de Lagôa, *Fernão de Magalhães: A sua vida e a sua viagem* (Lisboa, Seara Nova, 1938).

[67] Charles Nowell, *Magellan's Voyage Around the World: Three Contemporary Accounts* (Evanston: Northwestern University Press, 1962).

[68] Samuel Eliot Morison, *The European Discovery of America - The Southern Voyages* (New York: Oxford University Press, 1974).

[69] Tim Joyner, *Magellan* (Maine: International Marine, 1992).

9

A Great Ending to a Bad Start
Exiting the Magellan Strait (November 28th, 1520)

Even without the fleet's largest ship, Magellan was still determined to go forward with the mission, so he advanced the remaining trio of ships up the strait, likely to Carlos III Island (Fig. 9.1).[1] He knew the *San Antonio* deserters would portray him as a raging lunatic who would stop at nothing to achieve his goals, so it was now more important than ever to leave a paper trail painting his actions in a more favourable light. Magellan convened a second council to decide what to do, now demanding written answers from all officers. The letter he sent them seems as skewed as the accounts of the *San Antonio* deserters. Magellan described himself as "*a man who never refuses the opinion of anyone*", urging his officers not to let fear interfere with their oath of providing counsel, as "*what happened in Port San Julián was done in the service of His Majesty and for the security of his fleet*".[2]

San Martín suggested exploring the channels for another three weeks, while they have "*the flower of summer in hand*", and then "*Your Grace will judge whether to return to Spain, for from then on, the days will grow shorter and the weather more severe than it is now*".[3] Dreading Magellan would insist on travelling further south in search of a passage instead of returning home, San Martín adds a question—"*If we are now faced with such labour and risk, what will it be like at 70° or 75°, to which Your Grace said he would go in*

---

[1] Tim Joyner, *Magellan* (Maine: International Marine, 1992), on p. 331, note 37.

[2] João de Barros, *Década terceira da Ásia de João de Barros* (Lisboa: Jorge Rodriguez, 1628), on Livro Quinto, Capítulo VIII, f.143.

[3] Idem.

R. Gaspar, *The Revolution of Magellan*, https://doi.org/10.1007/978-3-032-10797-8_9

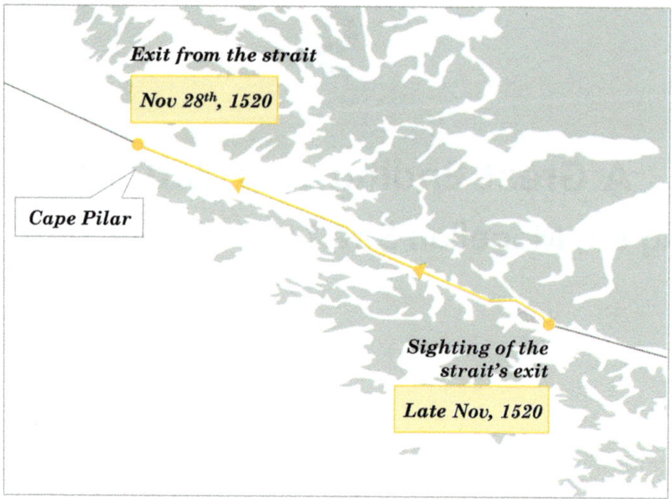

Fig. 9.1 Fleet route[4]

search of the Moluccas?".[5] A legitimate question indeed, as these extreme latitudes would put the fleet well within the Antarctic Circle. Only San Martín's response has reached our days, but it seems doubtful the other officers dared to be more assertive than the astronomer. After hearing his council, Magellan announces his decision to continue exploring the strait.

As the crew from the longboat had reported, after Carlos III Island the strait continues northwest as far as the eye can see, getting progressively wider. After a few days of sailing, on November 28th, the fleet finally sights the tip of land which marks the end of the strait, and bestows it with the rare honour of two names: *Cabo Hermoso* (Cape Beautiful) and *Cabo Deseado* (Cape Desired).[6] It was an historical moment, proving the Atlantic Ocean was indeed not landlocked to the west, and creating a flurry of economic opportunities.

After the double-named cape, today known as Cape Pilar, there was nothing but a vast expanse of water, prompting Pigafetta to make an astonishingly accurate observation:—*"When we left that strait, if we had sailed*

---

[4] Fleet route adapted from Tomás Mazón Serrano ("Mapas", Ruta Elcano, June 19, 2024, https://rutaelcano.com/mapas/); map adapted from Natural Earth (1:10 m large scale land data, version 5.1.1, https://www.naturalearthdata.com/downloads/).

[5] João de Barros, *Década terceira da Ásia* […], on f.144.

[6] Francisco Albo, *Crónicas de la Primera Vuelta al Mundo, según sus Protagonistas*, transliterated by Cristóbal Bernal, 2016, on p. 391.

continuously westward we would have circumnavigated the world without finding other land than the Cape of the Eleven Thousand Virgins [Cape Dungeness]".[7]

9.1 Did the Strait of Magellan Strait Ever Get Much Use?

> **The short answer**
>
> *Hopes were high for the strait, but it soon became clear that crossing it was a far more dangerous affair than Magellan's no-incident pioneering passage suggested. Follow-up expeditions were met with impenetrable weather, and many ships either sunk or cowered back. The Strait of Magellan developed such a bad reputation that for a long time ships preferred to go round Cape Horn— discovered in 1616 and even more hazardous. In the 1800's, ships hunting whales, seals and sea lions ventured back into the strait, and the navigational data they collected made its way to more dependable charts and weather forecasts. In the early 1900's, the Strait of Magellan finally became less feared and more popular than Cape Horn. It was a short-lived victory, though. In 1914 the Panama Canal opened, once again plunging the strait into obscurity.*

The news of a passage connecting the Atlantic and the Pacific generated considerable excitement in Europe, prompting a flurry of follow-up expeditions. The first to follow on the heels of Magellan was Francisco García Jofre de Loaísa, who left Spain in 1525 at the helm of a sizeable fleet of seven ships and 450 men. At the entrance of the strait, the armada faced gale force winds and was forced to retreat to the Atlantic. In the process, the *San Lesmes* caravel, commanded by Francisco de Hoces, was blown southward to a latitude of 55° and reported seeing the "end of land" before scrambling back north to regroup with the fleet. The expedition eventually made it through the strait, but not without losing the *Sancti Spiritus* in the process.[8]

Over the next 30 years, Spain launched six more attempts to cross the Strait of Magellan (Fig. 9.2, Cabot to Ladrillero), both westward (seeking the Moluccas) and eastward (departing from the American west coast, with the goal of exploring and controlling the southern coastlines). In three of the expeditions no ship was able to cross the strait, while in the remaining three only a third to half of the ships managed to do so. The discouraging success rate made it clear Magellan had benefited from unusually calm conditions.

---

[7] Antonio Pigafetta, *The first voyage around the world 1519–1522*, edited by Theodore Cachey Jr. (Toronto: University of Toronto Press, 2007), on p. 25.

[8] José Maria Moreno Madrid and Henrique Leitão, *Atravessando a porta do Pacífico* (Lisbon: By the Book, 2020), on p. 27, 28.

The less lucky ones who followed were battered by the strong winds and currents which typically plague the strait for most of the year. Williwaws—sudden blasts of wind that come hurtling down snow-capped mountains and glaciers—are particularly terrifying.[9]

The Spanish Crown was not the only one interested in the opportunities offered by the Strait of Magellan. While there are rumours of earlier clandestine expeditions, the first country to openly defy Spain's claim to the strait was Britain, in 1577, with a fleet led by Francis Drake. Drake also benefitted from atypical weather and crossed the strait in a mere 16 days, a record which would take centuries to match.[11] However, unlike Magellan, Drake was greeted by fierce storms after venturing into the Pacific. His remaining three-ship flotilla was summarily disbanded. The *Marigold* sunk and the *Elizabeth* re-entered the strait and headed back to England. Drake's *Golden Hind* (originally the *Pelican,* renamed after entering the Pacific) was pushed far down the coastline before managing to claw its way up again.[12]

Due to this unintended detour, Drake is often credited with the discovery of the *Drake Passage*—the stretch of water below the Americas, where the Pacific meets the Atlantic—but there is no evidence he ever sighted open seas, let alone crossed them back into the Atlantic. It seems more likely that

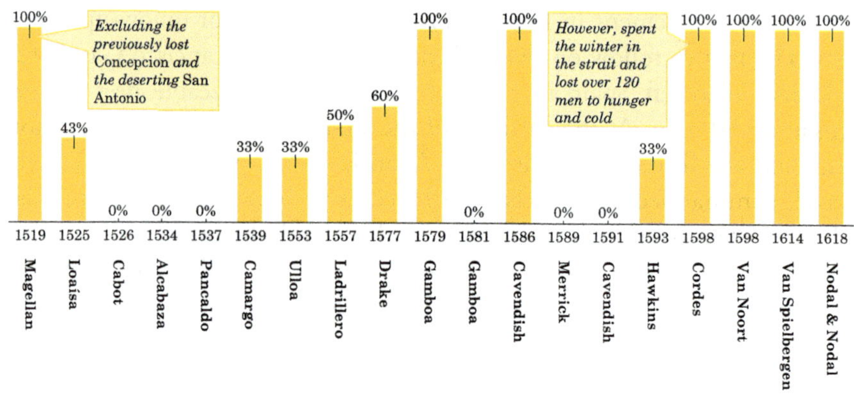

Fig. 9.2 Proportion of ships in each fleet which successfully crossed the Strait of Magellan, for the first 100 years after Magellan's passage[10]

---

[9] Tim Joyner, *Magellan* (Maine: International Marine, 1992), on p. 165.

[10] Adapted from: José Maria Moreno Madrid and Henrique Leitão, *Atravessando a porta do Pacífico* [...], on p. 27–33; James Burney, *Chronological history of the voyages and discoveries in the South Sea or Pacific Ocean* (London: Luke Hansard, 1806), on Volume 2, p. 191.

[11] Mateo Martinic, *Historia del Estrecho de Magallanes* (Santiago de Chile: Editorial Andres Bello, 1977), on p. 67.

[12] José Maria Moreno Madrid and Henrique Leitão, *Atravessando a porta do Pacífico* [...], on p. 30.

Drake's pilots noted the steep eastward slant of the coastline and inferred, perhaps building upon Hoces' earlier observations, that the "end of land" was near.[13]

The definite confirmation that the South American continent ended shortly after the Strait of Magellan came only in 1616, with Willem Schouten. At the turn on the century, the Netherlands had become the second European country to openly challenge Spain's claims in the region. The first three Dutch expeditions (two in 1598 and one in 1614, Fig. 9.2) had targeted the strait, but Schouten aimed instead at exploring further south, where the observations of Hoces and Drake hinted at the existence of an alternative way into the Pacific. He indeed found one, and named the southernmost point of the passage Cape Horn (in honour of Hoorn, the city where Schouten had been born[14]).

Rounding Cape Horn is notoriously treacherous, as the absence of land masses at this latitude allow gales to whirl around the globe unabated. Many modern-day mariners consider it the *Everest* of sailing but, in the seventeenth century, it was considered safer than the strait. One can hardly fault the conclusion, given the Strait of Magellan's spotty record. The use of Cape Horn for transit between the Atlantic and the Pacific plunged the strait into 200 years of obscurity, a period in which it was left to filibusters, pirates, and smugglers, plus the occasional scientific expedition.

The strait's popularity made a slow comeback in the 1800s Fig. 9.3), with the arrival of whalers and other ships hunting seals and sea lions. The geographical and meteorological information gathered in these trips trickled into charts, which allowed other vessels to negotiate the passage with far less risk. The development of oceanic steamships also contributed to an increase in traffic, as these liners—less dependent on the whims of winds and currents—preferred the winding but shorter passage to the more open but longer Cape Horn route. Around the turn of the century, the Strait of Magellan finally became more used than Cape Horn.[15]

Further north however, a competing engineering endeavour of byzantine proportions was firmly underway (Fig. 9.4). The idea of blazing a canal through Central America was an old one. Not long after Balboa first stood in the narrow Isthmus of Panama, in 1513, the Spaniards had toyed with the idea of a man-made passage to the Pacific. However, in 1534, after conducting a land survey, the governor of Panama disappointingly informed

---

13 Mateo Martinic, *Historia del Estrecho de Magallanes* [...], on p. 57, 68.

14 "Willem Schouten", *Britannica*, retrieved November 2024, https://www.britannica.com/biography/Willem-Schouten.

15 Mateo Martinic, *Historia del Estrecho de Magallanes* [...], on p. 150–153.

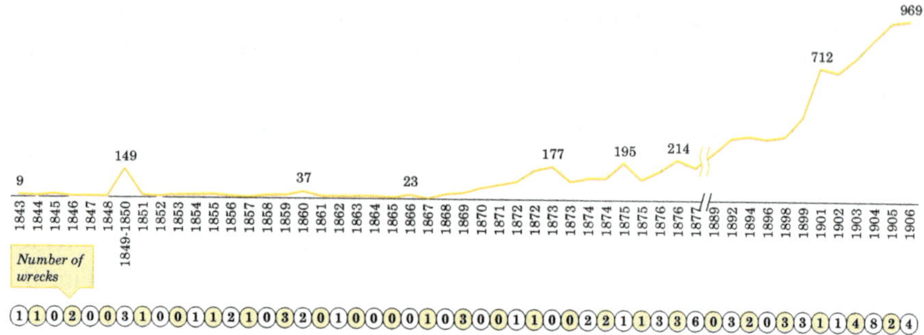

Fig. 9.3 Yearly maritime traffic and number of wrecks in the Strait of Magellan[16]

King Charles V such a canal was beyond the capabilities of mankind.[17] With sixteenth century resources it was certainly so. Even at the narrowest part of the isthmus, a canal required cutting through more than 60 km of steep rocky terrain, thick jungles, and deep swamps.

Over the next centuries, several plans for a canal were brought forward but never quite materialised. The first actual construction attempt started in the late nineteenth century, led by France under a concession provided by Columbia (which, at the time, included Panama). After the successful construction of the Egyptian Suez Canal, Ferdinand de Lesseps, a French diplomat, secured the funds necessary to replicate the project in Panama. However, it soon became clear this would be a far more complex undertaking. In Egypt, the engineers had found almost ideal conditions for a man-made passage, as the land was close to sea-level and easy to dig, and the weather was hot but dry. In Panama however, the terrain was tortuous and mostly made of rock, and the tropical humidity invited disease. They also needed to divert the course of the Chagres, a raging river which would otherwise flood the waterways during the rainy season.[18]

Construction began enthusiastically in 1881, but difficulties quickly mounted. Cutting through rock was hard work, and workers had to continuously widen the canal to reduce the angle of its slopes and avoid landslides. Every month hundreds of workers succumbed to yellow fever and malaria (thought to be caused by "noxious vapours" released from the ground, rather

---

[16] Mateo Martinic, *Historia del Estrecho de Magallanes* [...], on p. 164, 168, 169, 253, 254, 255.

[17] Thomas Cutler, "The Oregon and the Panama Canal", *US. Naval Institute Proceedings*, Vol. 145/11/ 1,401, 2019, https://www.usni.org/magazines/proceedings/2019/november/oregon-and-panama-canal.

[18] David McCullough, *The path between the seas—The creation of the Panama Canal: 1870–1914* (New York: History Book Club, 2002), on p. 76.

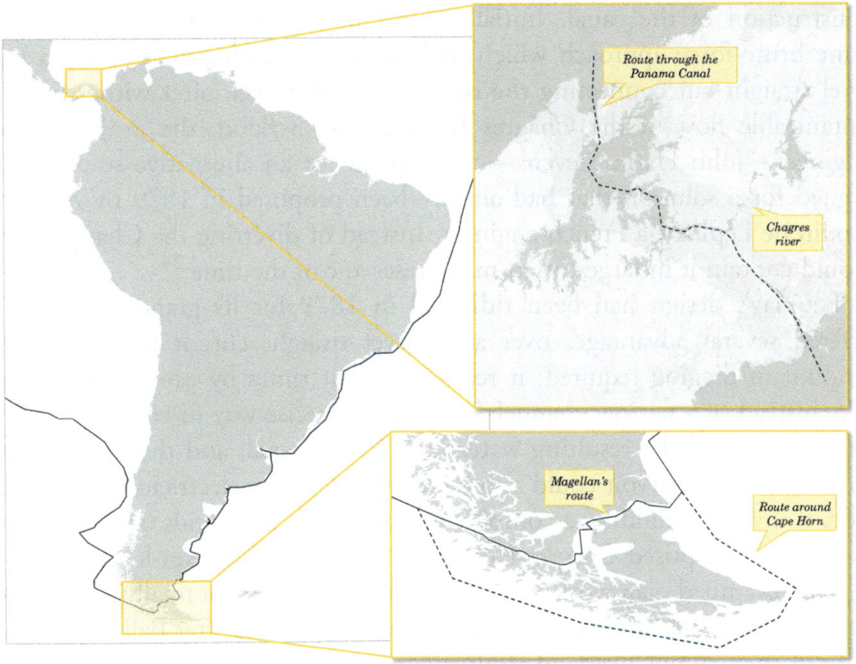

Fig. 9.4 The Panama Canal (top right) allows ships to shortcut through Central America instead of circumventing South America, greatly reducing travel distances compared to the Strait of Magellan or Cape Horn (bottom right)[19]

than by mosquito bites[20]). The project was halted in 1889, leaving behind 22,200 workers dead[21] and 800,000 small investors out of pocket.[22]

The French made a second half-hearted construction attempt in 1894. This time, with limited funds available, their main goal was merely to keep complying with the conditions of the concession while seeking a buyer for the project's infrastructure and resources. In the early twentieth century the United States agreed to buy the French operation, at the same time backing the Panamanian rebels who sought independence from Colombia.[23]

In 1904, following a successful rebellion, the US signed a new concession deal with the newly independent nation of Panama and took over the

---

[19] Magellan's route adapted from Tomás Mazón Serrano, "Mapas", Ruta Elcano, June 19, 2024, https://rutaelcano.com/mapas/; map adapted from Natural Earth (1:10 m large scale land data, version 5.1.1, https://www.naturalearthdata.com/downloads/).

[20] David McCullough, *The path between the seas* [...], on p. 142.

[21] Ralph Avery, *America's triumph at Panama* (Chicago: Regan Printing House, 1913), on p. 58.

[22] Deborah Cadbury, *Seven wonders of the industrial world* (London: Fourth Estate, 2003), on p. 262.

[23] David McCullough, *The path between the seas* [...], on p. 361–386.

construction of the canal. Initially, the North Americans insisted on the same brute-force approach which had spelled doom for the French: a sea-level straight cut connecting the two oceans. However, after witnessing the untameable flow of the Chagres River during a flood, the project's chief engineer—John Frank Stevens—campaigned for an alternative strategy. He argued for a solution that had already been proposed in 1879 by Adolphe Godin de Lépinay, a French engineer. Instead of diverting the Chagres, they would contain it in largest man-made reservoir of the time.[24]

Lépinay's design had been ridiculed in 1879 for its grandiosity, but it offered several advantages over a sea-level straight cut: it minimised the amount of digging required; it reduced transit times by providing a broad lake instead of a narrow channel; it offered a precise way to control the flow of the river and the resulting water level in the canal; and the dam needed to contain the reservoir could also be used to generate electricity. On the flip side, it involved building a complex set of locks at both ends of the passage, as the newly baptised Chagres Lake would sit 26 m above sea-level.[25]

The revamped plan worked. In 1914 the Panama Canal finally opened, 30 years after the first construction attempts and 400 years after Balboa had first crossed the isthmus overland. Immediately, traffic in the Strait of Magellan fell precipitously, relegating it back to an exile in the remote south. Today, the strait continues to play a negligible role in international maritime traffic, but its history and landscapes attract the more intrepid travellers.

9.2   What Happened to the Campfires in Tierra del Fuego?

> ***The short answer***
>
> *Apart from sporadic contacts, the Fuegians enjoyed nearly 300 years between Magellan's expedition and the arrival en masse of Europeans, due to the lack of interest in settling Tierra del Fuego. Things only changed in the 1800's, with an influx of whalers into the strait and the use of the surrounding barren landscapes for sheep farming. Similarly to what had happened in the rest of the Americas, acculturation, miscegenation, conflict, and disease quickly dwindled the number of natives. Today, less than 10,000 Chileans and Argentineans self-identify as one of the original Fuegian peoples.*

---

[24] David McCullough, *The path between the seas* [...], on p. 307.
[25] Idem, on p. 80, 81.

The smokestacks Magellan's crew could see rising from the shorelines of the strait were campfires of at least three peoples known to have inhabited this harsh corner of the globe: the Alacalufe (endonym *Kawésqar*), the Yahgan (*Yámana*), and the Ona (*Selk'nam*) (Fig. 9.5).

The Ona lived to the east of the strait, at *Isla Grande*, the largest of the *Tierra del Fuego* islands, today split roughly down the middle between Chile and Argentina. They were closely related to the Tehuelche of Patagonia, sharing the same nomadic way of life, language, and physical appearance—men were tall and lean, while women were shorter and stouter.[27]

The Yahgan roamed the lands and canals of the lower portion of the strait, all the way down to Cape Horn, making them the world's southernmost people. Despite being nomads like the Ona, the Yahgan were seafarers which relied on the sea for travelling and hunting, moving from island to island on bark canoes just big enough for a family and their dog. Also, they did not resemble the Ona at all: in lieu of long and angular proportions, the Yahgan were short and stocky. To the considerable shock of outsiders, they endured the freezing weather with the barest of clothes, with children often wandering around stark naked. While they could have fashioned clothes from fur seals and otters, these would do them little good in a region where is rains

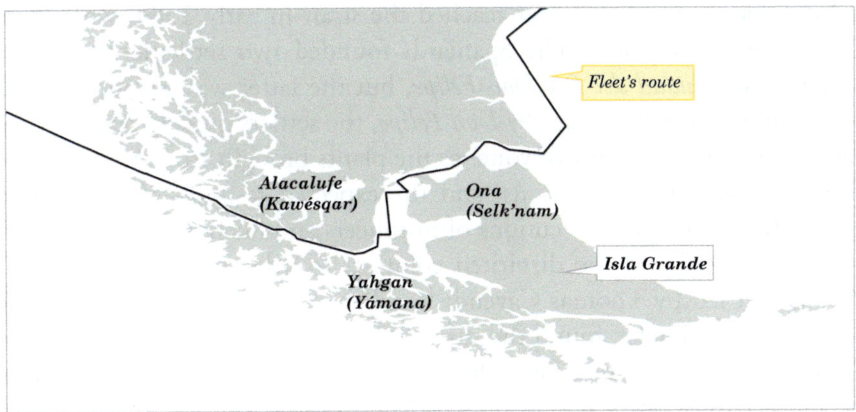

Fig. 9.5 Overview of the peoples inhabiting Tierra del Fuego at the time of the expedition[26]

---

[26] Fleet route adapted from Tomás Mazón Serrano ("Mapas", Ruta Elcano, June 19, 2024, https://rutaelcano.com/mapas/); map adapted from Natural Earth (1:10 m large scale land data, version 5.1.1, https://www.naturalearthdata.com/downloads/); tribal footprints adapted from Dallas Murphy, *Rounding the Horn: Being the story of williwaws and windjammers, Drake, Darwin, murdered missionaries and naked natives—a deck's-eye view of Cape Horn* (London: Phoenix, 2005), on p. 136.

[27] "Tehuelche", Britannica, retrieved February 2025, https://www.britannica.com/topic/Tehuelche-people.

nearly 300 days per year. Instead, they acclimated to these harsh conditions, developing fatter builds and a higher metabolism. Whenever possible, they would squat close to a fire—which they kept permanently alight, even on the canoes—conserving as much heat as possible.[28]

The western portion of the strait was inhabited by the *Kawésqar*, as they called themselves, but became more widely known as the Alacalufe ("mussel eaters"),[29] which was what the Yahgan called them. The two groups had the same physical appearance and habits but different languages, and their territories often overlapped in the southwestern parts regions of the strait.

Magellan's crew did not report any direct contact with natives along the strait. However, if they were able to see their campfires, the locals also saw their giant wind-propelled canoes, and perhaps elected to remain out of sight. No direct confrontation is reported until 1578, when Francis Drake killed Alacufes while docked in a cove in the western part of the strait.[30]

Colonial terrestrial expansion in South America halted in 1567, with Spanish settlers uninterested in going south of the Island of Chiloé (latitude 43°). However, Drake's and other British corsair raids in Chilean colonies prompted the Spanish Crown to order the fortification of the strait. In 1581, a gigantic fleet of 23 ships and 3,000 men—including 350 settlers and 400 soldiers to man the forts—left Spain, led by Gamboa. After several wrecks and desertions, a reduced fleet reached the strait in early 1584, with moral already dangerously low. The Spaniards founded two settlements, *Nombre de Jesús* and *Ciudad del Rey Don Felipe*, but the latter was abandoned after five months. All huddled at *Rey Don Felipe*, the settlers quickly realised there was not enough food for everyone, as the plants brought from Spain refused to grow in the unfamiliar soils, and the expected resupply ships failed to arrive. Hunger, disease, executions of mutineers and likely also conflicts with the residents reduced the dispirited colonists to a handful of survivors. In 1587 a fleet led by Thomas Cavendish, a British corsair following the footsteps of Francis Drake, found the survivors, taking one of them prisoner and leaving the others to their fate. One of those left behind somehow endured for another three years and was rescued by Andrew Merrick in 1590. After nearly a decade, the proud fleet of 3,000 strong had thus dwindled to two

---

[28] Dallas Murphy, *Rounding the Horn* [...], on p. 131–144.

[29] Joy McCann, *Wild Sea: A history of the Southern ocean* (Chicago: The University of Chicago Press, 2019), on p. 216.

[30] Mateo Martinic, *Historia del Estrecho de Magallanes* (Santiago de Chile: Editorial Andres Bello, 1977), on p. 62–67.

lonely survivors. The ruins left behind at *Rey Don Felipe* were renamed to Port Famine by Cavendish, a name that still stands today.[31]

The horrors of Port Famine cooled down Spain's interest in colonising the strait for a long time, once again limiting the contacts between Europeans and Fuegians to sporadic brawls with sailors. One such clash though, in 1599 with Van Noort's Dutch expedition, left behind a bloody trail of 40 dead Onas.[32]

The frequency of encounters increased steeply in the 1800s, with the arrival of whalers and ships hunting seals and sea lions.[33] The Yahgan and the Alacalufe, who depended on these marine mammals for subsistence, were the most affected. And, just as the other Native Americans,[34] the Fuegians were not immune to Europeans diseases. Together with malnutrition, illnesses such as smallpox and measles precipitously reduced their numbers. A young Charles Darwin, who crossed the strait aboard the *HMS Beagle* in 1832, had the following to say about the Yahgan he encountered:

> These poor wretches were stunted in their growth, their hideous faces bedaubed with white paint, their skins filthy and greasy, their hair entangled, their voices discordant, and their gestures violent. Viewing such men, one can hardly make oneself believe that they are fellow-creatures, and inhabitants of the same world. It is a common subject of conjecture what pleasure in life some of the lower animals can enjoy: how much more reasonably the same question may be asked with respect to these barbarians![35]

The landlubbing Ona were less affected by the increasing number of ships in the strait, but their fortune started to slowly turn in 1843. The young Republic of Chile, which had gained independence from Spain in 1818, took formal possession of the Strait of Magellan by establishing Fort Bulnes, very close to the gloomy ruins of Port Famine. Chile, much like the Spanish Empire 250 years before, was more interested in controlling marine traffic in the strait than in colonising the surrounding lands. Occupying *Isla Grande*—the home of the Ona and the vastest parcel of *Tierra del Fuego*—was a particularly fiddly affair, as Argentina also held claims over this area.[36]

---

31 Idem, on p. 113–119.

32 Mateo Martinic, *Historia del Estrecho de Magallanes* [...], on p. 74.

33 See Did the Strait of Magellan Strait Ever Get Much Use?

34 See What Happened to the Natives Encountered by the Fleet?

35 Charles Darwin, *The Voyage of the Beagle* (New York, New American Library, 1988), on p. 183, 184.

36 Mateo Martinic, "Recordando a un imperio pastoril: La Sociedad Explotadora de Tierra del Fuego (1893–1973)", *Magallania*, 39(1), 2011, p. 5–32.

That changed with the Boundary Treaty of 1881, where Chile and Argentina settled their border disputes. *Isla Grande* was cut vertically, with Argentina holding on to the eastern half and Chile keeping the western part, together with the strait. Shortly after, in 1894, Chile granted several sheep farming concessions, the largest one to the *Sociedad Explotadora de Tierra del Fuego* (Company for the Exploitation of *Tierra del Fuego*). The business grew quickly, and at one point managed over 3,000,000 hectares on both sides of the border i.e., most of *Isla Grande*.[37]

The company's sheep quickly displaced the resident population of guanacos. Deprived of sustenance and not privy to the concept of private property, the Ona hunted the sheep—the "white guanacos", as they called them. This did not sit well with the ranchers, who in turn hunted the Ona. Militia were brought in and allegedly paid for each killed Ona.[38] News of the matter rippled across society, causing consternation. Mauricio Braun, one of the *Sociedad Explotadora*'s shareholders, justified their actions in a letter to a fellow owner:

> This matter of the Indians is rather unpleasant, but what to do, we need to extirpate them from Tierra del Fuego and take them all to Dawson Island[39]

The Chilean government had allowed Italian Salesians to set up a mission at Dawson Island to harbour exiled Fuegians. The ranchers would pay one pound sterling for each native brought to the mission, an amount that would readily prove insufficient to ensure the safeguard the hundreds of Ona hurtled across the channel. Most of them, uprooted from their homes and weakened by malnutrition, died from smallpox and measles.[40] In 1895, a census conducted by the Chilean government had counted 1,500 Ona. Fifteen years later, around 100 were left.[41]

---

[37] Idem.

[38] Anne Chapman, *Drama and Power in a Hunting Society: The Selk'nam of Tierra Del Fuego* (Cambridge: Cambridge University Press, 1982), on p. 11.

[39] Mateo Martinic, *Panorama de la colonizacion em Tierra del Fuego entre 1881 y 1900* (Punta Arenas: Instituto de la Patagonia, 1973), on p. 42.

[40] Idem, on p. 43.

[41] Idem, on p. 57, 63.

Today, the nomadic way of life of all these peoples is long gone. Their languages are also extinct (for the Yahgan) or on the brink of extinction (for the Ona and Alacalufe). Nevertheless, their DNA and some aspects of their cultures live on in the bodies and souls of Chileans and Argentineans who self-identify as Fuegian descendants (~4,000 Onas, ~3,500 Alacufe, and ~1,600 Yahgan).[42]

---

42 "Chile Censos 2017", Instituto Nacional de Estadísticas, retrieved July 2024, https://www.ine.gob.cl/estadisticas/sociales/censos-de-poblacion-y-vivienda; "Argentina Censo 2022", Instituto Nacional de Estadística y Censos, retrieved July 2024, https://www.indec.gob.ar/indec/web/Nivel4-Tema-2-41-165.

10

Crossing Earth's Largest Ocean

From the exit of the Magellan Strait (November 28th, 1520) to Guam (March 6th, 1521)

Magellan and his crew were now part of the select group of European ever to lay eyes on the waters which extended to the west of the Americas. Of course, they had no way of knowing that the expanse they were now entering was the same *Mar del Sur* (Southern Sea) Balboa had christened a few years back.[1] Instead, they called it *Pacifico* (Peaceful)[2] (Fig. 10.1).

In stark contrast with the explorations of the South American coastline, this new body of water did not greet them with any storms,[3] nor did it hurl them into the jagged Chilean coastline. However, the crew's initial good spirits slowly faded away, as days trickled down into weeks, and weeks turned into months. Already short on rations when they left the strait, each passing turn of the onboard sand clocks led them closer to despair.

Twice they sighted land, but in both instances were unable to take on fresh supplies. The first sighting was on January 24th, roughly two months after leaving the strait.[4] The small atoll, surrounded by waters too deep to drop anchor, was likely Puka-Puka, currently part of French Polynesia.[5] Not long after, on February 4th,[6] they sighted *Isla de los Tiburones* (Shark Island,

---

[1] See How Did Magellan Convince the Spanish King?

[2] Antonio Pigafetta in H.E.J. Stanley, *The first voyage round the world, by Magellan. Translated from the accounts of Pigafetta, and other contemporary writers* (London: Hakluyt Society, 1874), on p. 65.

[3] Idem.

[4] Francisco Albo in Cristóbal Bernal, *Crónicas de la Primera Vuelta al Mundo, según sus Protagonistas*, 2016, on p. 394.

[5] Tim Joyner, *Magellan* (Maine: International Marine, 1992), on p. 169, citing Denucé, Lagôa, and Morison.

[6] Francisco Albo in Cristóbal Bernal, *Crónicas de la Primera Vuelta al Mundo* […], on p. 395.

R. Gaspar, *The Revolution of Magellan*, https://doi.org/10.1007/978-3-032-10797-8_10

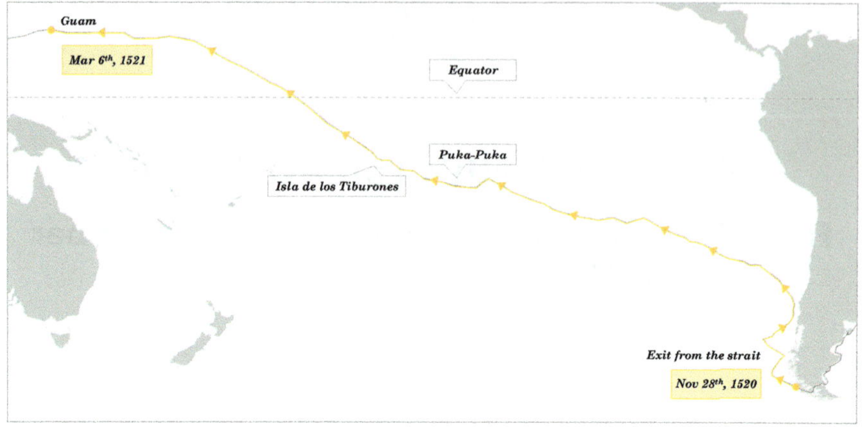

Fig. 10.1 Fleet route[7]

perhaps present-day Flint Island[8]) but, after a brief inspection, also deemed it unworthy of a stop.

Pigafetta's descriptions of those days, while perhaps exaggerated by the Italian's customary literary flair, provide a glimpse into the crew's struggles. About the food, the Italian wrote that "*we ate biscuit, which was no longer biscuit, but fistfuls of powder swarming with worms, for they had eaten the better part (it stank strongly of rat urine)*[9]". The rodents were no longer a nuisance but a delicacy—"*Rats were sold for one-half ducado[10] apiece, and even then we could not get them[11]*". Food was washed down with "*yellow water which had been putrid for many days*", as the wine supplies were long gone. Pigafetta also reported that "*we also ate some ox hides that covered the top of the main yard to prevent the yard from chafing the shrouds, and which had become exceedingly hard because of the sun, rain, and wind. We left them in the sea for four or five days, and then placed them for a few moments on top of the embers, and thus we ate them*", fulfilling Magellan's previous ominous assertion.[12]

---

[7] Fleet route adapted from Tomás Mazón Serrano ("Mapas", Ruta Elcano, June 19, 2024, https://rut aelcano.com/mapas/); map adapted from Natural Earth (1:10 m large scale land data, version 5.1.1, https://www.naturalearthdata.com/downloads/).

[8] Tomás Mazón Serrano, "Mapas" […], on *Isla de los Tiburones (hoy de Flint)* log entry.

[9] Antonio Pigafetta, *The first voyage around the world 1519–1522*, edited by Theodore Cachey Jr. (Toronto: University of Toronto Press, 2007), on p. 24.

[10] Half a ducat was nearly 190 marevedies, or about 15% of the monthly salary of a sailor. One can only conjecture if rats were indeed such high-prized delicacies, or if Pigafetta was exaggerating.

[11] Antonio Pigafetta, *The first voyage around the world* […], on p. 24.

[12] In the middle of the strait, faced with his crew's pleas to return to Spain, Magellan reportedly stated that "*even if they had to eat the leather on the ships' yards, he would still go on, to discover what he had promised to the king*" (see A Bad Ending to a Great Start).

At least 25 men fell ill during the crossing, most of them from scurvy.[13] This debilitating disease typically sets in after 60 to 90 days of a vitamin C deficient diet and moves through four increasingly agonising stages. In the first stage, the affected sailors feel fatigue and their muscles and joints ache. The second stage adds swollen gums and loose teeth. In the third stage, the gums become putrid and smell of rotten flesh, and body skin turns gangrenous and bleeds. The fourth and final stage ends in excruciating death, normally caused by haemorrhaging in the brain or heart.[14]

The effects of scurvy can be reversed, even in advanced stages, by consuming vitamin C. However, the cure eluded scientists for centuries, with some estimates blaming scurvy for killing over two million sailors between the fifteenth and eighteenth centuries.[15] Only in 1929 did Albert Szent-Györgyi, a Hungarian biochemist, demonstrate that a compound he called hexuronic acid had antiscorbutic properties.[16] Hexuronic acid was later renamed to ascorbic acid (meaning "anti-scurvy"), and is today more commonly known as vitamin C. Before the cure for scurvy was formally discovered, sailors and other expeditioners attempted a myriad of different treatments. Some of these were utterly useless (such as seawater, vinegar, and vitriol) but the understanding it was helpful to consume fresh foods (oranges and lemons in particular) was slowly established over the centuries.[17] One of the first documented uses of citrus fruits to cure scurvy goes back to Vasco da Gama's 1497

---

[13] Antonio Pigafetta, *The first voyage around the world* [...], on p. 24.

[14] Jason Mayberry, "A Timeline of Scurvy", LEDA at Harvard Law School, 2004, https://dash.har vard.edu/bitstream/handle/1/8852139/Mayberry.html.

[15] Kenneth Carpenter, *The history of scurvy and vitamin C* (Cambridge: Cambridge University Press, 1986), on p. 253, citing Carey McCord.

[16] Albert Szent-Györgyi, *"Observations on the functions of peroxidase systems and the chemistry of the adrenal cortex: Description of a new carbohydrate derivative"*, Biochemical Journal, 22(6), 1928, p. 1387–1409, on p. 1408, 1409, doi.org/10.1042/bj0221387.

[17] Jonathan Lamb, *Scurvy: The Disease of Discovery* (Princeton: Princeton University Press, 2016), on p. 1–26.

discovery of the sea route from Portugal to India. When the fleet stopped in modern-day Kenya, scurvy-ridden sailors recovered six days after eating oranges.[18] Ten years later, Pedro Álvares Cabral also noted the refreshments (likely orange or lemon juice) offered by locals were effective against the disease.[19]

There is no evidence suggesting Magellan had access to the findings of his fellow countrymen. Even if he did, he would not find any citrus fruits along the fleet's route, as they had not yet been introduced in South America. However, knowingly or not, Magellan and other officers may have been able to keep healthier than the sailors by consuming the quince jelly which was part of their improved provisions,[20] as it contains some vitamin C.

More than 20 men succumbed from scurvy.[21] Among them was "Paulo", one of the two Tehuelche they had abducted in Port San Julián,[22] who was likely not high up the pecking order for the shrinking rations. A disproportionate number of casualties were travelling in the *Victoria*,[23] suggesting its crew may have followed a distinct diet. While still in the strait, the sailors had access to some food sources rich in vitamin C, such as the stalks of celery-like Apium plants[24] abundantly present in Bahía Fortescue, fish roe, and seal liver (Fig. 10.2). Sailors who ate enough of these foods in the strait would be less likely to die from scurvy during the three-month Pacific crossing, as the disease typically requires two to three months of not consuming vitamin C. One can only conjecture, but could the *Victoria's* crew have had less access to these food sources, or have been fussier with what they ate?

On March 6th, more than three months after leaving the Strait, the fleet sighted land again. This time there was little doubt they could gather some provisions here, as on approach the ships were swarmed by swift canoes

---

[18] Kumaravel Rajakumar, "Infantile scurvy: a historical perspective", *Pediatrics*, 108(4), 2001, doi.org/10.1542/peds.108.4.e76.

[19] Cristina Gurgel and Rachel Lewinsohn, "Medicine on board the caravels - sixteenth century", *Cadernos de História da Ciência*, 6(2), 2010, p. 105–120, on p. 108.

[20] See What Did 16th Century Sailors Eat?

[21] Antonio Pigafetta, *The first voyage around the world* [...], on p. 24.

[22] The other Tehuelche died aboard the deserting San Antonio, see A Bad Ending to a Great Start

[23] José Toribio Medina, *El descubrimiento del Océano Pacífico: Vasco Núñez de Balboa, Fernando de Magallanes y sus compañeros* (Santiago de Chile: Imprenta Universitaria, 1920), on p. CCLXVIII, CCLX.

[24] Teresa Carvalho, "Documenting the tropical natural world in the account of Antonio Pigafetta", *Magallánica: Revista de Historia Moderna*, 7(13), 2020, p. 288–314, on p. 298.

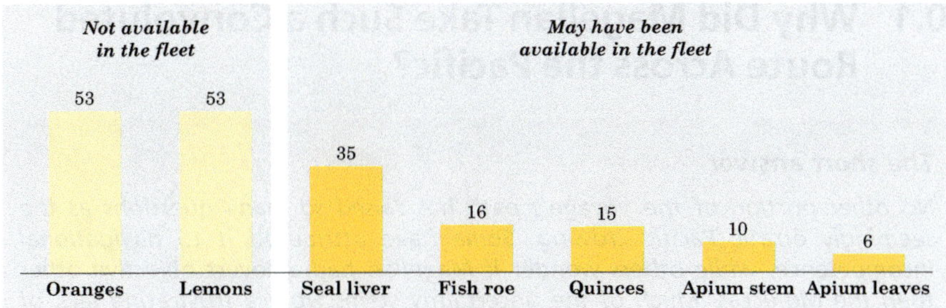

Fig. 10.2 Comparison of vitamin C content (in mg of vitamin C per 100 g serving) in different foods[25]

packed with Chamorros, the natives of modern-day Guam. They had just completed the first documented crossing of the Pacific,[26] but Pigafetta was convinced it could not be repeated: "*Had not God and His blessed mother given us such good weather we would all have died of hunger in that exceedingly vast sea. In truth I believe no such voyage will ever be repeated*".[27]

For once, Pigafetta was not exaggerating. The most common view of our globe, centred in the Greenwich prime meridian, hides away the Pacific Ocean. When peeking around the back though, its scale becomes immediately apparent. The Pacific takes up almost a third of the Earth's surface and is larger than all landmasses combined. Magellan's expedition managed to cross it on the first try, with no charts of the region nor prior knowledge of its prevailing winds, aboard battered ships, and with scurvy-debilitated sailors who could barely stand up. Plus, they did it immediately after surviving a winter in Patagonia and navigating their way through the Strait. Short of space exploration, no other expedition since came even close to traversing some 18,500 km[28] of unchartered territory, let alone on the first try.

---

[25] Adapted from: "USDA National Nutrient Database for Standard Reference Release 28", USDA, 2015, https://ods.od.nih.gov/pubs/usdandb/VitaminC-Content.pdf; Joseph Geraci and Thomas Smith, "Vitamin C in the Diet of Inuit Hunters From Holman, Northwest Territories", *Artic*, 32(2), 1979, p. 135–139, on p. 137; Bahare Salehi et al., "Apium Plants: Beyond Simple Food and Phytopharmacological Applications", *Applied Sciences*, 9(17), 2019, https://doi.org/10.3390/app9173547, on p. 8.

[26] But not necessarily the first crossing, see Who Were the First to Cross the Pacific?

[27] Antonio Pigafetta, *The first voyage around the world* […], on p. 25.

[28] Tomás Mazón Serrano, "La primera vuelta al mundo", Ruta Elcano, assessed November 2024, https://rutaelcano.com/la-primera-vuelta-al-mundo/.

10.1 Why Did Magellan Take Such a Convoluted Route Across the Pacific?

> **The short answer**
>
> *No other portion of the voyage's path has raised so many questions as the seemingly erratic Pacific crossing. Some have attributed it to navigational incompetence, while others wonder if Magellan had a covert objective other than the Moluccas. Much of the uncertainty stems from a frustrating lack of first-hand reports, as Pigafetta and other crew members wrote very little about the passage. However, other pieces of evidence can still be used to dismiss much of the intrigue: Magellan and his pilots were not surprised by the size of the Pacific nor were they hopelessly lost; and most of the course decisions are consistent with the typical wind conditions for this part of the world.*

The shortest route from the Strait of Magellan to Guam is about 14,400 km, some 4,000 km short of the distance travelled by the fleet. This is not a sizeable difference for a sail ship, bound by the whims of winds. However, the path itself seems odd, as the three ships first went north up the Chilean coastline before turning sharply east, and then continued travelling northeast well past the equator, where they knew the Moluccas stood. Were they lost, taken aback by the size of the Pacific? Or maybe they were not looking for the Moluccas but something else? And how come did they only spot two islands out of the thousands scattered throughout the Pacific? Were Magellan's spotters near-sighted or dozing off in the crow's nest?[29]

The latter question is the easiest to address. Sighting land is as difficult today as it was in the sixteenth century. Even with perfect visibility, the curvature of the Earth hides any low island or atoll further than 30 kms.[30] Indeed, finding unchartered islands was—and would still be, if there were any uncharted islands left—the nautical equivalent of finding a needle in a haystack.

However, not all islands in the vicinity of the fleet's route were small and flat. One such exception was the Juan Fernandez archipelago, off the coast of Chile. According to Albo's logbook, the ships passed about 115 km northeast of the archipelago, which is less than the ~135 km minimum distance necessary to spot the El Yunque peak (Fig. 10.3). Perhaps the fleet was further

---

[29] A small platform near the top of the main mast (Fig. 5.6).

[30] Assuming a crow's nest at 25 m, islands with maximum elevation of 5 m, and considering the effects of refraction.

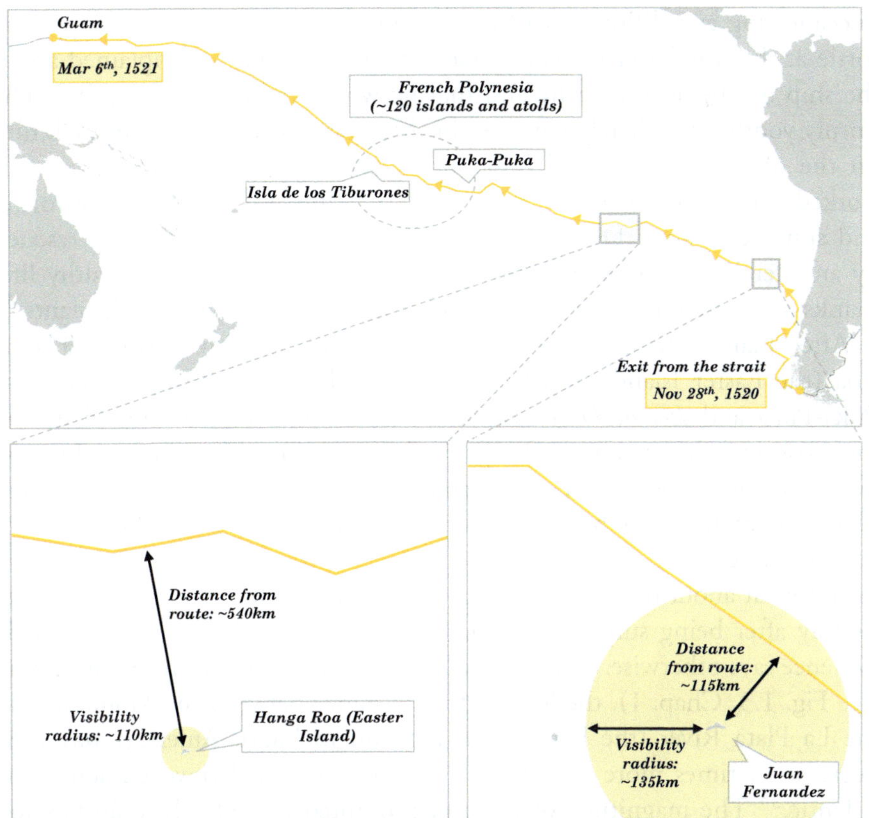

Fig. 10.3 The fleet's route across the Pacific, showing how close the ships would need to pass to spot Juan Fernandez and Hanga Roa, the two islands with the highest peaks in the vicinity of the route[33]

away than what Albo's logbook suggests,[31] or the day was cloudy[32]? Whatever happened it was unfortunate, as Juan Fernandez would have allowed Magellan to restock food supplies, as an impertinent officer would learn nearly 200 years later.

Alexander Selkirk, who inspired the Robinson Crusoe novel, was an early eighteenth century Scottish privateer and Royal Navy officer. As part of a

---

[31] Sixteenth century pilots were able to measure latitude with an accuracy of about 0.5° (see What Navigation Techniques Were Available to Magellan?) or 60 km, enough to place the fleet further away than the minimum 135 km viewing distance.

[32] Or maybe the spotters were indeed distracted, although the likelihood of this happening simultaneously on the three ships seems slim.

[33] Juan Fernandez's highest peak is El Yunque (915 m) and Hanga Roa's is the Terevaka Volcano (507 m). Viewing distances assume a spotter atop a 25 m high mast and take into account the effects of refraction.

buccaneering expedition aboard the *Cinque Ports*, and during a resupplying sortie at the uninhabited Juan Fernandez islands, Selkirk complained about the ship not being seaworthy and that he would rather be left behind. The unruly youth immediately regretted his hasty words, as the captain took him on the offer and marooned him on the desert island. In retrospect, this worked out for the best for Selkirk, as the ship was indeed unseaworthy and sunk soon after. During his four-year seclusion—before being rescued by another English privateer ship—Selkirk lived a solitary but healthy life, thanks to the island's ample supplies of water, fish, game, and edible plants.[34]

After Juan Fernandez, the ships sailed some 540 km north of Hanga Roa (the Easter Island), too far to spot the Terevaka Volcano. Except for Puka-Puka and *Isla de los Tiburones*—the only two islands spotted during the crossing—the remaining Pacific landmasses were either too small or too distant to be visible. So then, Magellan was unlucky in missing the Juan Fernandez archipelago, but it is no surprise thousands of other Pacific islands went unnoticed.

But what about the fleet's winding route? Some believe the expedition lost its way after being surprised by the true scale of the Pacific. The available evidence says otherwise. In the chart Magellan presented to the Spanish king (see Fig. 1.3, Chap. 1), the longitudinal distance between the Moluccas and the La Plata River (the last plotted location in South America) was 178° degrees, 13 times more than the 14° degrees they had travelled across the Atlantic.[35] The magnitude of the task prompted the pilot Estevão Gomes, while still in the strait, to urge Magellan to turn back to Spain, as there was still a "great gulf to pass".[36]

But why did the fleet first sail straight up, if they knew their objective lied well to the west? The three ships which had made it through the Strait—the *Trinidad*, the *Concepción*, and the *Victoria*—were all 85+ ton naus with sails that allowed for the transport of heavy cargo but limited the ability to travel upwind.[37] Accustomed to the Portuguese trading routes in the Atlantic and Indian oceans, Magellan knew it was better to follow trade winds rather than force the heavy ships onto a more direct route. Once out of the Strait, he may have guessed—or at least hoped—the south Pacific trade winds followed

---

[34] Woodes Rogers, *A Cruising Voyage Round the World: First to the South Sea, Thence to the East Indies, and Homewards by the Cape of Good Hope* (London: A. Bell, 1712), on p. 91–96.

[35] From the Canaries to Cape Frio, in Brazil.

[36] José Toribio Medina, *El descubrimiento del Océano Pacífico: Vasco Núñez de Balboa, Fernando de Magallanes y sus compañeros* (Santiago de Chile: Imprenta Universitaria, 1920), on p. CCXLIX.

[37] See How Advanced Were Magellan's Ships?

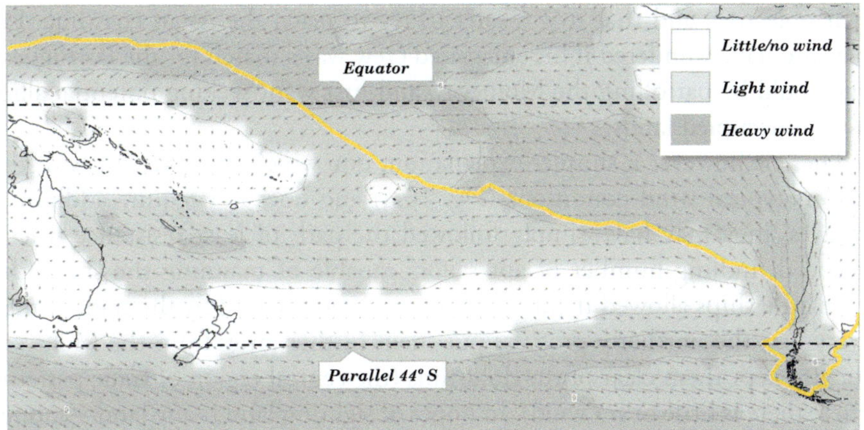

Fig. 10.4 The fleet's route across the Pacific overlaid with the typical wind conditions for January (midpoint of the three-month passage)[38]

the same clockwise pattern he had experienced before, allowing the fleet to eventually turn west. And he was indeed right (Fig. 10.4).

But while trade wind patterns can explain the route taken above latitude 44° S or so, it is harder to understand how Magellan was able to easily exit the strait. As several follow-up expeditions would learn the hard way, the Pacific in this region is usually not peaceful at all, as exceedingly strong head winds unceremoniously push the ships onto the Chilean cliffs, back into the strait, or around Cape Horn. Some believe the uncommon calm conditions enjoyed by Magellan were caused by the tail end of an El Niño event.[39]

Lastly, it is worthwhile to investigate the final portion of the passage, made above the equator. Why did the fleet travel into the northern hemisphere, if they knew the Moluccas straddled the equator? Many theories have been put forward as potential explanations. Magellan may have assumed there would not be enough food supplies in the tiny Moluccas and decided to first feed his starving crew elsewhere.[40] Or perhaps he feared they had already overshot

---

[38] January wind speed and direction averaged for the 1991 to 2020 period ("IRI/LDEO Climate Data Library", International Research Institute for Climate and Society, 2024, https://iridl.ldeo.col umbia.edu/); fleet route adapted from Tomás Mazón Serrano, "Mapas", Ruta Elcano, June 19, 2024, https://rutaelcano.com/mapas/.

[39] Scott M. Fitzpatrick and Richard Callaghan, "Magellan's Crossing of the Pacific: Using Computer Simulations to Examine Oceanographic Effects on One of the World's Greatest Voyages", *The Journal of Pacific History*, 43(2), 2008, p. 145–165, on p. 165.

[40] This is indeed what one of his pilots declared: *"They ran on until they reached the line, when Fernan de Magalhaes said that now they were in the neighbourhood of Maluco [Moluccas], as he had information that there were no provisions at Maluco, he said that he would go in a northerly direction as far as ten or twelve degrees"* (the Genoese pilot in H.E.J. Stanley, *The first voyage round the world, by*

their objective, and that it was safer to try to reach the Asian coastline. He may also have wondered if his emaciated crew was a match for the Portuguese who could be patrolling the Moluccas, and first went seeking potential allies on non-occupied lands.

There is however a simpler explanation, one which is consistent with the region's typical wind conditions (Fig. 10.4) and does not require additional motives: they may have simply followed the path of least resistance. Faced with a dwindling number of apt sailors capable of manning the ships, the three little ships allowed themselves to be pushed forward by the trade winds. Every day, the northerly progress put them further away from the Moluccas but also kept them clear of the doldrums (the central region of trade patterns where there is little or no wind), in hopes of sighting land—any land—sooner rather than later.

Magellan. Translated from the accounts of Pigafetta, and other contemporary writers, London, Hakluyt Society, 1874, on p. 9).

11

Bad News

From Guam (March 6th, 1521) to Homonhon, Philippines (March 25th, 1521)

After more than three months without seeing a single living soul, the fleet's first interaction with Pacific Islanders would not leave the best initial impression. After swarming around the ships in their nimble canoes, the native Polynesian Chamorros climbed aboard and, before the hungry and sick crew could do anything about it, took off with anything that was not nailed down to the deck. Among the pilfered items was the *Trinidad's* skiff, which was being trailed behind the ship (Fig. 11.1).[1]

To avoid any more unexpected visitors under the cover of darkness, the fleet retreated and spent the night offshore.[2] The next day, they returned with a landing party of more than 40 armed men to retrieve the skiff. It seems Magellan also wanted to teach the Chamorros a hard lesson, as his crew inflicted punishment far worse than the crime. Armed with crossbows and fighting only against stones and spears, the landing party reportedly killed seven villagers, burned dozens of huts and canoes, and snatched fresh supplies and water from the village.[3] To add insult to injury, they named the place *Isla de los Ladrones* (Island of the Thieves), a name which stuck for 140 years.[4]

Despite the rough start, a few Chamorros cautiously came back, trading fresh goods and indicating that a larger quantity of supplies could be sourced

---

[1] Francisco Albo in Cristóbal Bernal, *Crónicas de la Primera Vuelta al Mundo, según sus Protagonistas*, 2016, on p. 397.

[2] Jean Denucé, *Magellan: la question des Moluques et la première circumnavigation du globe* (Brussels: Hayez, 1911), on p. 306.

[3] Antonio Pigafetta, *The first voyage around the world 1519–1522*, edited by Theodore Cachey Jr. (Toronto: University of Toronto Press, 2007), on p. 27.

[4] Paul Carano and Pedro C. Sanchez, *A Complete History of Guam* (Rutland: Tittle, 1964), on p. 62.

R. Gaspar, *The Revolution of Magellan*, https://doi.org/10.1007/978-3-032-10797-8_11

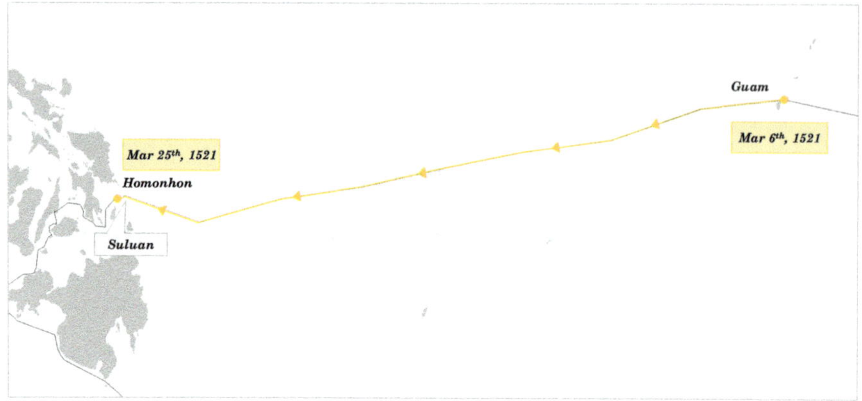

Fig. 11.1 Fleet route[5]

for an island further west.[6] Following the tip, the fleet set sail and, on March 16th, sighted the mountains of the large Samar Island, part of modern-day Philippines. Reefs prevented a landing, so the crew explored instead a set of smaller islands, spending less than a day at Suluan before deciding to anchor at Homonhon.[7] The fresh goods sourced from Guam had come too late for a few of the men afflicted with scurvy, who died on their way there. For the others still recovering, Magellan ordered tents to be raised on firm land.[8]

A couple of days after they had arrived, a pirogue carrying nine men approached them. Initially cautious, Magellan relaxed after realising the visitors were friendly and willing to trade. After exchanging everything they had on their boat—a much-welcomed jug of palm wine, fish, bananas, and coconuts—the natives vowed to return shortly with additional supplies.[9]

When they did, Magellan took the opportunity to invite them aboard. He showed them cloves and other spices, which the locals confirmed grew in the region.[10] This was the expedition's second defining moment: after successfully

---

⁵ Fleet route adapted from Tomás Mazón Serrano ("Mapas", Ruta Elcano, June 19, 2024, https://rutaelcano.com/mapas/); map adapted from Natural Earth (1:10 m large scale land data, version 5.1.1, https://www.naturalearthdata.com/downloads/).

⁶ José Toribio Medina, *El descubrimiento del Océano Pacífico: Vasco Núñez de Balboa, Fernando de Magallanes y sus compañeros* (Santiago de Chile: Imprenta Universitaria, 1920), on p. CCLXXII.

⁷ Francisco Albo in Cristóbal Bernal, *Crónicas de la Primera Vuelta al Mundo* [...], on p. 398.

⁸ Antonio Pigafetta, *The first voyage around the world* [...], on p. 30.

⁹ Idem.

¹⁰ Idem, on p. 31.

finding and navigating the passage between the Atlantic and the Pacific, they now had confirmation they had reached the Indies by sailing west. It had been anything but easy, but Magellan had succeeded where Christopher Columbus did not.

There was a problem, though. During their brief stay at Suluan, the astronomer San Martín had made a longitude measurement, and the results were not what Magellan had hoped for: they seemed to be 189° west of the Tordesillas demarcation line, already 9° inside Portugal's hemisphere.[11] If true it was catastrophic, as it meant that these islands, and most likely the nearby Moluccas, belonged to Portugal, not Spain.

11.1 Were San Martín's Longitude Measurements Accurate?

The short answer

San Martín's longitude measurement at Suluan was indeed accurate, and so was the one he had previously made at Port San Julián. These results have baffled historians for centuries, as the method he used, called lunar distances, was still 200 years away from becoming reliable. NASA's latest astronomical tables, used for space exploration, have helped to clear the mystery. As it turns out, San Martín's accurate results were due to fortuitous cancellations of large errors, and the astronomer had no way of knowing they were correct. Luck seemingly ran out on his other four other attempts at determining longitude, which did not produce usable results. However, it would be rash to summarily dismiss San Martín's work. He was an exceedingly rare combination of a capable astronomer and a knowledgeable mariner, but had to work within the limitations of 16th century science.

From the onset, Magellan knew the success of his expedition hinged on a single number: the longitude of the Moluccas. If measurements revealed the islands were on the Portuguese hemisphere, the feat of finding a passage through the Americas and surviving the Pacific crossing would be of little value.

---

[11] Rolando Laguarda Trías, "Las longitudes geográficas de la membranza de Magallanes y del primer viaje de circunnavegación", *Actas do II Colóquio Luso-Espanhol de História Ultramarina*, 1975, p. 135–174, on p. 170.

In an alternative scenario where Portugal and Spain had split the planet horizontally instead of vertically, Magellan's expedition probably would have not happened, as the latitude of the Moluccas was not up for debate. However, longitude was far more difficult to measure accurately in the sixteenth century.

Determining latitude (i.e., the ship's position north or south of the equator) was already an integral part of navigation and the fleet's pilots routinely did it with an error of less than 1°, by using a marine astrolabe or quadrant to take the altitude of the Sun or another star.[12] The accuracy of these measurements was 1:1, meaning that a 1° error in the altitude of the star (or in the tables used to convert it to latitude) turned into a 1° error in latitude.

Longitude, on the other hand, was not used for navigation. Longitude pinpoints the east–west position of the ship relative to a reference meridian (Greenwich nowadays, Seville for Magellan's voyage). This reference meridian rotates as the day goes by, completing a full 360° revolution every 24 h, or 15° per hour. Longitude is thus equivalent to the time difference between two places e.g., Athens is 30° east of Seville, so the sun shines over the Acropolis two hours before it looms over Seville's Cathedral. The crux of the issue is precisely this 1:15 conversion factor between time and longitude: to measure the latter with an accuracy of 1°, one needs to measure the former within a mere four minutes.

None of Magellan's pilots knew how to measure longitude, except San Martín. The Spaniard, who was both a pilot and an astronomer, placed Suluan Island 9° west of the Tordesillas anti-meridian (132° east of Seville), an error of less than 2°, or 8 min. And this was not his best measurement: while the fleet was still in South America, San Martín placed the ill-fated San Julián port[13] 61° west of Seville, an error of a mere 0.6°, or 3 min.[14]

It is worthwhile to put these results into context. Sixteenth century European attempts to measure longitude overseas routinely had errors one or two orders of magnitude above San Martín's best results, often requiring multiple observations to narrow down to reasonable results. Even within Europe, with access to a greater pool of experienced astronomers and better equipment, errors of more than 10° (40 min) were not uncommon (Fig. 11.2). How did San Martín accomplish such a feat?

---

[12] See What Navigation Techniques Were Available to Magellan?

[13] The place where Magellan's crew rebelled, see The Uprising.

[14] Romeu Gaspar, "San Martín's accurate longitude measurements on Magellan's circumnavigation: luck or mastery?", *Journal of Navigation*, 2024, p. 1–27, https://doi.org/10.1017/S037346332400016X, on p. 4.

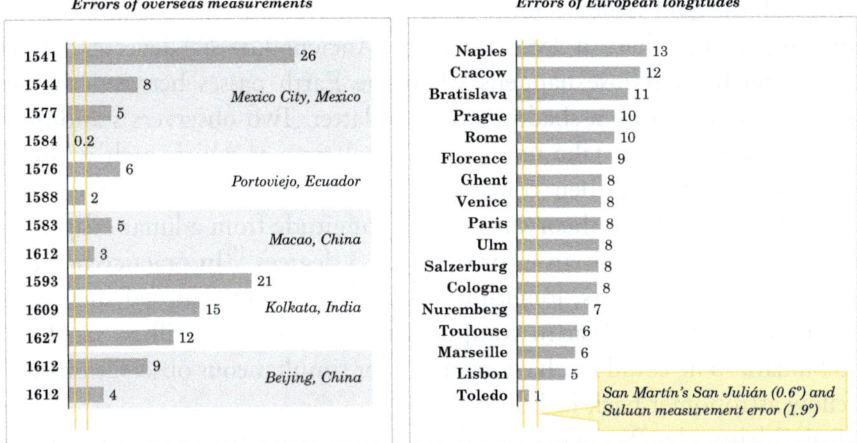

Fig. 11.2 Examples of errors of 16th- and 17th-century European longitude readings (all errors in degrees of longitude)[15]

Today, the most straightforward way to measure longitude, bar a GPS,[16] is with a watch. At departure, a sailor going from Seville to Athens will set a timepiece at solar noon. Every day, the mariner can track easterly progress by checking the time difference between his watch (which provides him with the reference time at Seville) and solar noon at its current position (i.e., local time). When the difference between reference time and local time reaches two hours, the seafarer has covered the 30° longitude separating Athens from Seville.

Of course, San Martín did not have a watch, nor did his peers. The time-keeping device available aboard sixteenth century ships was the hourglass, usually in 30 min and 2 h varieties. Even if San Martín had a perfectly calibrated hourglass (with just the right amount of sand and bottleneck diameter to precisely time two hours) and somehow immune to the ship's constant swaying, by the time the fleet reached Suluan it would have been turned 6,516 times (543 days, 12 turns per day). To keep within the 8 min error of his measurement, each turn of the hourglass could not be off by more than 6 s,[17] an impossible feat for the cabin boys responsible for the task.[18]

---

[15] Adapted from: W.G.L. Randles, "Portuguese and Spanish attempts to measure longitude in the sixteenth century", *Vistas in Astronomy*, 28, 1985, p. 235–241, on p. 235–238; Abraham Zacut, *Almanach perpetuu[m] exactissime nuper eme[n]datu[m] omniu[m] celi motuum* (Venice: Pietro Liechtenstein, 1502), on Tab. I.

[16] *Global Positioning System*, which has become ubiquitous in everything from ships to smartphones.

[17] 8 min divided by the square root of 6,516.

[18] See What Was the Life of a 16th Century Sailor Like?

Instead, the gold standard for measuring longitude in the sixteenth century (and long before that, at least since the Ancient Greeks) leveraged lunar eclipses. Such an eclipse happens when the Earth passes between the sun and the moon, casting a shadow onto the latter. Two observers standing in different locations can take note of the local time at which each witnesses the lunar eclipse, and then convene to convert the time difference into a longitude difference. In theory, measuring longitude from a lunar eclipse was reasonably accurate, with an error of about 3 degrees.[19] In practice however, and as Fig. 11.2 suggests, longitude readings seldomly achieved such accuracy. In addition, San Martín computed longitudes while still at San Julián and Suluan, so he could not have arranged for simultaneous observations with a fellow astronomer back in Europe.

Rui Faleiro, the Portuguese mathematician and astronomer with whom Magellan initially partnered,[20] had prepared a treatise that described three alternative methods of measuring longitude. Faleiro's preferred method, which took up a sizeable portion of the treatise, was devoted to the chimera of using magnetic declination to measure longitude, a method first described by the Portuguese pilot João de Lisboa in 1514.[21] Knowing that magnetic declination was zero in the Canary Islands, Faleiro postulated that it varied linearly along the east–west axis and could thus be used to measure longitude. While sometimes there is indeed a local correlation between magnetic declination and longitude (for instance, Portuguese pilots exploited it to estimate how far they were from the Cape of Good Hope[22]), the relationship breaks down across long distances as magnetic declination does not vary linearly.

Faleiro devoted little more than a couple of sentences to each one of the other two methods, both astronomical. Despite the frugality, these techniques held one major advantage over lunar eclipses: they replaced the second observer by an almanac, allowing San Martín to measure longitude by himself and without waiting to get back to Europe. These almanacs, which predicted the future position of celestial bodies (the Sun, the Moon, plus visible planets

---

[19] Romeu Gaspar, "San Martín's accurate longitude measurements on Magellan's circumnavigation" [...], on p. 8.

[20] See How Did Magellan Convince the Spanish King?

[21] And later included in his *Livro de Marinharia* (Lisbon: Arquivo Nacional da Torre do Tombo, PT/TT/CRT/166, 1560).

[22] José Malhão Pereira, "The bearing compass and the magnetic variation in Portuguese navigation techniques", Estudos da História da Náutica e das Navegações de Alto-Mar (Lisboa: Edições Culturais da Marinha, 2013), on p. 116.

and stars) were popular among astronomers and astrologers, and San Martín had taken a couple of them aboard.[23]

Both methods were based on the motion of the Moon, but one used the satellite's ecliptic latitude (the Moon's position north or south of the ecliptic, which is the plane where the Earth orbits the Sun, Fig. 11.3), while the other one relied on ecliptic longitude (the Moon's position east or west of the vernal equinox[24]). In a letter addressed to Magellan,[25] San Martín swiftly dismissed using ecliptic latitude to compute longitude. He had good reasons to do so: while in theory he could measure the Moon's local ecliptic latitude and compare it with the ecliptic latitude in Seville given by his almanacs, in practice the method required a precision far beyond what was possible with sixteenth century instruments and almanacs (the Moon's ecliptic latitude changes 0.03° per hour,[26] so a 1° error in measurement would turn into an enormous 500° error in longitude[27]).

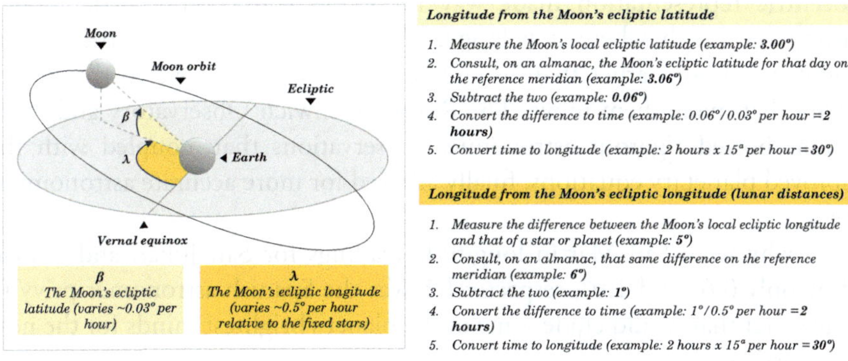

Fig. 11.3 Steps required to obtain the longitude of a place from the Moon's ecliptic latitude and ecliptic longitude

---

[23] Romeu Gaspar, "San Martín's accurate longitude measurements on Magellan's circumnavigation" […], on Appendix A, p. 21–23.

[24] The start of Spring in the Northern Hemisphere (between March 19th and 21st), used as the reference point for ecliptic longitude.

[25] Fernão Lopes de Castanheda, *História do descobrimento & conquista da Índia pelos portugueses* (Lisbon: Typ. Rollandiana, 1833), on p. 13.

[26] The Moon's orbital plane around the Earth is tilted 5.1° relative to the ecliptic. On every revolution of the Moon around the Earth, its ecliptic latitude will thus vary between +5.1° and −5.1°, changing 0.75° per day, or 0.03° per hour.

[27] 1° divided by 0.03° per hour times 15° per hour.

Not all was lost though, as San Martín was more enthusiastic about Faleiro's final proposal—the one based on ecliptic longitude—stating that *"if there is another way to measure the difference from east to west I do not know it"*.[28] The procedure, which would later become known as the *lunar distances* method, had been described more thoroughly a few years before by a German mathematician called Johannes Werner.[29] While measuring the Moon's ecliptic longitude is less demanding than measuring ecliptic latitude (a 1:30 error[30] instead of 1:500), it was still well beyond the reach of sixteenth century science. In fact, it would take more than 200 years to turn the lunar distances method into something more than a throw of the dice. Multiple European naval powers, including Spain, France, and Britain, launched sizeable prizes in exchange for a practical way to measure longitude aboard, attracting contributions from some of the Renaissance's brightest minds. Eventually, the octant and the sextant replaced the quadrant and the astrolabe, substantially reducing the error of angle measurements. Ptolemy's geocentric representation made way for Copernicus's heliocentric model, putting the Earth where it belonged. Kepler determined that the planets moved in eclipses rather than in circles, and Newton discovered that it was gravity that made them move. The Royal Greenwich Observatory and other stations hosted rigorous astronomical observations that, coupled with the improved planetary equations, finally allowed for more accurate astronomical tables.

So, why were San Martín longitude readings for San Julián and Suluan off by only 0.6° and 1.9°, respectively? Was the Spanish astronomer privy to some secret that would elude some of the planet's brightest minds for the next 200 years?

---

[28] Fernão Lopes de Castanheda, *História do descobrimento & conquista da Índia pelos portugueses* [...], on p. 13.

[29] Johannes Werner in *The quest for longitud: the proceedings of the Longitude symposium*, edited by William Andrewes and translated by Andrea Murschel (Cambridge: Harvard University Press), on Appendix C, p. 385.

[30] The Moon's ecliptic longitude changes approximately 0.5° per hour, so a 1° error in measuring means a 30° error in longitude (1° divided by 0.5° per hours times 15° degrees per hour).

As it turns out, these were not his only longitude readings during the expedition. San Martín had also placed Rio de Janeiro 269° west of Seville (a much less impressive error of 232°) and there are records of at least three other measurements, the results of which have unfortunately not survived.[31] So, the plot thickens: How did he achieve such extraordinarily accurate measurements at San Julián and Suluan, but was off by so much at Rio de Janeiro? And were the three lost measurements good or bad?

These questions have remained shrouded in mystery for the last 500 years. Recently, with the help of NASA's latest astronomical tables,[32] the veil has been lifted. These tables, designed for space exploration, are so precise that one can rewind the clock 500 years, see the same sky San Martín witnessed during his observations, and simulate his measurements (Table 11.1).

The results are fascinating. At Rio de Janeiro, San Martín never stood a chance, due to the poor accuracy of his almanacs (based, as all other sixteenth century almanacs, on Ptolemy's old equations of planetary motion[34]). Still, the Spanish astronomer misplaced Rio by over 230°, considerably more than the expected 28° error. San Martín was acutely aware the result made no sense (after all, it placed Rio de Janeiro more than halfway around the globe, in the middle of the Indian Ocean), and blamed the almanacs for the blunder. There was however one larger source of inaccuracy: San Martín himself. His

Table 11.1 Simulated results of San Martín's longitude measurement attempts, compared with the true longitude of each location and with the astronomer's own result (when known). All values are relative to the Seville reference meridian (a minus sign indicates a position west of Seville, a plus sign east of Seville)[33]

Date	Location	True longitude	San Martín's measurement	Simulated measurement
17/12/1519	Rio de Janeiro	−37.2°	−268.8°	−8.8 ± 0.0°
01/02/1520	La Plata River	−50.2°	Lost	−168.8 ± 11.5°
24/02/1520	Gulf of San Matias	−58.6°	Lost	+33.5 ± 11.2°
17/04/1520	**San Julián**	**−61.6°**	**−61.0°**	**−61.7 ± 2.2°**
23/12/1520	Pacific Ocean	−70.2°	Lost	+13.2 ± 10.6°
16/03/1521	**Suluan Island**	**+131.9°**	**+130.0°**	**+128.5 ± 11.3°**

---

[31] Romeu Gaspar, "San Martín's accurate longitude measurements on Magellan's circumnavigation" […], on p. 4.

[32] Ryan Park et al., "The JPL Planetary and Lunar Ephemerides DE440 and DE441", *The Astronomical Journal*, 161(3), 2021, p. 1–15, https://doi.org/10.3847/1538-3881/abd414.

[33] Romeu Gaspar, "San Martín's accurate longitude measurements on Magellan's circumnavigation" […], on p. 19, Table 3.

[34] Owen Gingerich, *The Book Nobody Read: Chasing the Revolutions of Nicolaus Copernicus*, (New York: Walker Books, 2004).

Table 11.2 Main sources of error on San Martín's San Julián and Suluan Island longitude measurements (a minus sign indicates an westward error, and a plus sign an eastward one)[37]

Error	Error description	San Julián	Suluan Island
Parallax	**Error of not accounting for parallax**	**−21.6°**	**−28.2°**
Equations	**Errors in the almanac's tabulated values**	**+19.8°**	**+17.4°**
Geometric	Error of not reducing measurements to ecliptic longitude	0°	+3.9°
Interpolation	Error of interpolating motion from discrete table entries	+0.2°	+1.8°
Meridian	Error in the true longitude of the reference meridian	+1.7°	+1.7°
Refraction	Error of not accounting for refraction	~0°	~0°
Instrument	Error of the measurement instruments	0°	±10.9°
Local time	Error in determining local time	±2.2°	±2.2°
Total	**Total compounded error**	**+0.1 ± 2.2°**	**−3.5 ± 11.3°**

notes[35] show how a calculation misstep (a minus sign in lieu of a plus one) threw the results off. The lapse is disarmingly easy to make, particularly for a Northern Hemisphere astronomer making his first observations in the Southern Hemisphere, and not quite used to seeing the sky flipped around.

At San Julián and Suluan, the simulations confirm that San Martín could have indeed arrived to accurate longitude measurements using the lunar distances method. But how? Perhaps his almanacs were accurate for the days in which the observations were made? Not quite. What happened instead is that other sources of inaccuracy offset the errors of the almanacs (Table 11.2). Most notably, San Martín did not take the Moon's parallax into account,[36] generating an error just large enough to cancel the almanac error, allowing the astronomer to fortuitously arrive to accurate longitude readings.

---

[35] San Martín's original notes were lost, and only parts of them have reached our days via second-hand reports. One of those reports, penned by Herrera, included a faithful transcription of the astronomer's calculations at Rio de Janeiro (Antonio Herrera, *Historia general de los hechos de los castellanos en las islas i Tierra Firme del Mar Oceano*, Decada II, Libro IX, Madrid: En la emplenta Real, 1601, on p. 132).

[36] Parallax is the apparent shift in position of a nearby object relative to a distant point when the viewpoint changes. For instance, if one puts a finger in front of a remote mountain and then moves the head up and down (or left and right), the finger will no longer be aligned with the mountain. The same thing happens with the Moon and the stars, as the former is nearby and the latter faraway. Without correcting for parallax, San Martín's readings of the position of the Moon were not comparable to those provided by his almanacs.

[37] Romeu Gaspar, "San Martín's accurate longitude measurements on Magellan's circumnavigation" […], on p. 15, 17.

As for the lost longitude readings taken at the La Plata River, the Gulf of San Matias, and in the Pacific Ocean (shortly after existing the strait), the simulations suggest the results were wrong by at least 80° (Table 11.1).

So, at Rio de Janeiro San Martín made a calculation misstep, his uncannily accurate results at San Julián and Suluan were the result of a fortunate cancellation of large errors, and the three remaining lost measurements likely yielded poor results. It would be rash to dismiss his work though, as he was a pilot first and an astronomer second.[38] This combination of skills was exceedingly rare in the sixteenth century (if not unheard of), and better-known royal pilots from the *Casa*, such as Amerigo Vespucci (San Martín's friend and mentor), did not possess it. In any case, even if Magellan had managed to convince a veteran astronomer to join the expedition, he too would be bound by the limitations of sixteenth century science.

Were San Martín and Magellan aware of the inescapable unreliability of 16th-century longitude measurements, or were they truly convinced that Suluan—and most probably the nearby Moluccas—were already in the Portuguese hemisphere? After the tribulations of his multiple measurements, it is hard to imagine San Martín putting enough faith in his results to categorically state they were a mere 9° away from the demarcation line. Even if he was unaware of the magnitude of parallax errors, he had complained several times about the inaccuracy of the almanacs. As for Magellan, there is little doubt about his respect for science, patent in the case he put forward to the Spanish king and on his insistence in having an astronomer aboard. However, there is also ample evidence of his stubbornness and willingness to pursue his goals at any cost. Would he throw in the towel because a number did not agree with him?

---

[38] Carmen Mena-García, "Know and master the astros. The pilot Andrés de San Martín and the Magallanes/Del Cano expedition", *Temas Americanistas*, 44, p. 197–231, on p. 214, 215.

<h1 style="text-align:center">12</h1>

Worse News

From Homonhon, Philippines (March 25th, 1521) to Mactan, Philippines (April 27th, 1521)

After about a week on Homonhon Island, most of the ailing crew members had recovered from the effects of malnutrition and scurvy. There was never a dull moment, though. On March 25th, shortly before the fleet was set for departure, Pigafetta fell into the water while fishing. The Italian gentleman, one of those rare cases of men who turned to the sea for adventure rather than by necessity,[1] did not know how to swim. Fortunately, he was able to catch a piece of rigging that was dangling in the water, and a few mariners heard his screams and came to the rescue.[2] Surprisingly, swimming skills were quite rare in sixteenth century Europe even amongst mariners, and Pigafetta was not the only crew member who did not know how to keep afloat.[3]

With a drenched but breathing Pigafetta back on board, the ships set sail to the west (Fig. 12.1). It seemed like an odd choice of direction as Magellan knew the Moluccas were near the equator (while San Martín's measurements may have made him question the island's longitude, he had no reason to doubt Portuguese intelligence regarding its latitude). Perhaps Magellan was searching for something else before making its way to the Moluccas? He may have learned from the natives that Homonhon was part of a large archipelago with its most populous city—Cebu—and its main trading port—Butuan—to the west. There, he would not only be able to source additional provisions but also establish local alliances. If San Martín's longitude measurements

---

[1] See What Was the Life of a 16th Century Sailor Like?

[2] Antonio Pigafetta in H.E.J. Stanley, *The first voyage round the world, by Magellan. Translated from the accounts of Pigafetta, and other contemporary writers* (London: Hakluyt Society, 1874), on p. 75.

[3] Richard Mandell, *Sport: A Cultural History* (New York: Columbia University Press, 1984), on p. 179, 180.

R. Gaspar, *The Revolution of Magellan*, https://doi.org/10.1007/978-3-032-10797-8_12

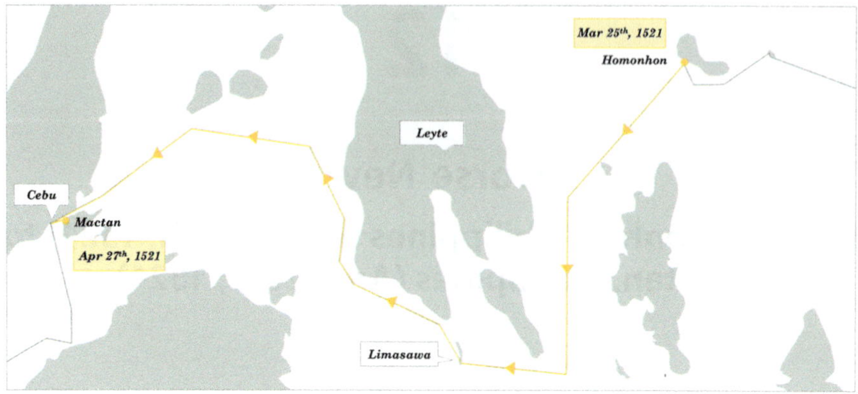

Fig. 12.1 Fleet route[6]

were right and this archipelago was already in Portugal's hemisphere, these local alliances could at least play a role in the forthcoming diplomatic clash between Spain and Portugal. And if San Martín's measurements were wrong and they were still on Spain's hemisphere, then the local alliances would make Magellan outrageously rich. The contract signed with the Spanish Crown was generous, stating Magellan would become governor of a few of the newly discovered territories and awarding him a share of the profits.[4] Magellan had good reason to suspect these *San Lázaro* islands (his name for the Philippines) were richer than the barren lands of Patagonia, as his crew had found traces of gold in a stream and the local chiefs used large gold armbands and earrings.[5]

After travelling southwest for about a day, the fleet sighted the eastern shore of the larger Leyte Island and turned south to circumvent it. On March 28th, while close to the Limasawa Island, a pirogue with eight locals approached the ships. Unsure about the visitors' intentions, they initially kept their guard up but became less worried after realising Enrique, a slave Magellan had bought in Malacca, was able to speak with them.[7] Enrique

---

[4] Of all the islands discovered by the fleet, the Crown would first choose six (likely the Moluccan islands, the prime objective of the trip). From the remaining Magellan could choose two, from which he would receive about 7% (one fifteenth) of the profits. From the rest he would receive 4% (one twenty-fifth) of the profits (Martin Fernandez de Navarrate, *Coleccion de los viajes y descubrimientos que hicieron por mar los españoles desde fines del siglo XV*, Tomo 4, Madrid, Imprenta Nacional, 1837, on p. 114, 118).

[5] Antonio Pigafetta, *The first voyage around the world 1519–1522*, edited by Theodore Cachey Jr. (Toronto: University of Toronto Press, 2007), on p. 32.

[6] Fleet route adapted from Tomás Mazón Serrano ("Mapas", Ruta Elcano, June 19, 2024, https://rut aelcano.com/mapas/); map adapted from Natural Earth (1:10 m large scale land data, version 5.1.1, https://www.naturalearthdata.com/downloads/).

[7] Antonio Pigafetta in H.E.J. Stanley, *The first voyage round the world, by Magellan.* [...], on p. 76.

and the locals likely communicated in Malay, the region's lingua franca[8], further reinforcing Magellan's belief they had reached the Indies. Aided by Enrique's language skills and Pigafetta's socialising ones, the Europeans gained the trust of the indigenous. The Limasawa Island was the hunting grounds for two brother chiefs—Colambu and Siaui—each one ruling over several surrounding islands and territories.[9] To demonstrate the value of establishing an alliance with Spain, Magellan showcased his three strengths to the two chiefs: valuable items to trade, a strong military force, and the blessing of God.

The locals were first showered with gifts (robes and caps of fine Turkish cloth for the chiefs, knifes and mirrors for their men[10]) and shown the multitude of articles for trade they had on stock. They responded with their own gifts, of which Magellan gratefully accepted the food but casually declined the gold and ginger, perhaps not to not draw attention to what Europeans valued.[11] The captain general then organised a visit to the flagship, where he fired the ship's cannons—utterly terrifying his guests—and organised a mock combat where three men armed with swords and daggers attacked, to no avail, another crew member in full amour.[12] While the natives marvelled at the imperviousness of the armour, Magellan boasted that each one of his three ships had two hundred of similarly armed men, a bold exaggeration. A few days later the captain followed up on the demonstration by volunteering his ships and men to dispose of their enemies. The chiefs acknowledged they were indeed at war with two neighbouring islands, but that the season for fighting had not yet arrived.[13]

On March 31st, Easter Sunday, Magellan had the perfect setting to showcase his third talent. He asked the chiefs for permission to celebrate mass ashore and invited them to attend. As Magellan and 50 of his men stepped into land in their finest attire, the ship's cannons fired in sign of peace. The chiefs were sprinkled with perfumed water and participated in the ceremony, raising their clasped hands to the body of Jesus Christ and mimicking the gestures of the Europeans. At the end of mass, Magellan offered them a tall cross with a crown of thorns, asking them to erect it at the island's highest

---

8 See Who Were the First to Circumnavigate the Globe?

9 Jean Denucé, *Magellan: la question des Moluques et la première circumnavigation du globe* (Brussels: Hayez, 1911), on p. 310.

10 Antonio Pigafetta in H.E.J. Stanley, *The first voyage round the world* […], on p. 77.

11 Or to downplay the value of the islands, so the Crown would not choose them as part of its first claim of six islands.

12 Antonio Pigafetta in H.E.J. Stanley, *The first voyage round the world* […], on p. 77.

13 Jean Denucé, *Magellan* […], on p. 311.

vantage point. This way, Enrique translated, all forthcoming Spanish ships would know they would find friends here.[14]

This was an effective strategy to form local alliances, but a time consuming one. Magellan may have thought it was more efficient to go straight to the archipelago's largest chieftain, rather than make the rounds through a multitude of small islands. The chiefs confirmed that Cebu was the region's largest city,[15] and Colambu even volunteered to serve as a pilot. On April 4th, the fleet set sail to the northwest, following Colambu's balangay (a lashed-lug boat commonly used for trading goods in the Philippines). Three days later, they arrived at Cebu. It was by far the busiest port they had encountered since crossing the Atlantic, with a shoreline filled with tightly packed houses on stilts and boats speeding up and down the canal. Yet, just as in the smaller islands, the indigenous ran away terrorised at the sound of the ship cannons.[16]

After the grand entrance, Cristovão Rebelo and Enrique were sent ashore to greet the local ruler.[17] Rebelo, Magellan's relative (some say his natural son[18]), was initially assigned as his servant but was now part of the captain general's inner circle of trust. Rebelo assured Humabon, Cebu's rajah, that the artillery shots were a sign of peace on behalf of King Charles V, the greatest king in the world. Fear not though, as they were there for trade and for friendship. Unperturbed, Humabon answered they were welcome to trade, provided they paid the port tariff as everybody else. Hearing that, Rebelo swiftly shifted to a more intimidating tone. They had come in peace, he retorted, but if war was what Humabon wanted they would oblige. A trader from Siam (modern-day Thailand) overheard the discussion and urged the Cebu rajah to be cautious. These were probably the same *farang*[19] pirates who had taken Calicut and Malacca by force, terrorising the natives and interrupting the traditional trading routes which had been in place for centuries.[20]

The Siamese trader had taken the Spaniards for Portuguese, who by this time already controlled the major trading routes in the Indian Ocean. In

---

[14] José Toribio Medina, *El descubrimiento del Océano Pacífico: Vasco Núñez de Balboa, Fernando de Magallanes y sus compañeros. Fernando de Magallanes* (Santiago de Chile: Imprenta Universitaria, 1920), on p. CCLXXIV, CCLXXV.

[15] Antonio Pigafetta in H.E.J. Stanley, *The first voyage round the world* [...], on p. 82.

[16] José Toribio Medina, *El descubrimiento del Océano Pacífico* [...], on p. CCLXXVI.

[17] Antonio Pigafetta in H.E.J. Stanley, *The first voyage round the world* [...], on p. 84.

[18] José Toribio Medina, *El descubrimiento del Océano Pacífico* [...], on p. CCCCXXVII.

[19] A Persian word which initially referred to the Franks, but evolved to encompass westerners in general.

[20] Antonio Pigafetta in H.E.J. Stanley, *The first voyage round the world* [...], on p. 84, 85.

a sense he was right, as Magellan had been part of the Portuguese war fleet which went to the Indies in 1505. The more than 20 war ships carrying thousands of soldiers and hundreds of gunners were no match for the Egyptian-Indian-Venetian forces that attempted to protect their trading routes. In a strategy which would be re-used by allied forces in the Pacific War, the Portuguese leveraged their superior naval power to conquer strategic ports between Gibraltar and India, and later down to Malacca.[21] Magellan knew this plan had been set in motion years before any war ship left the Lisbon port. Under the guise of discovery and peaceful trade, Vasco da Gama and other Portuguese explorers had first gathered intelligence and built fortified trading posts along the route, paving the way for the war ships that followed.

After hearing the Siamese trader's warning, Humabon told Rebelo he would discuss the matter with his advisors and get back with a decision. He also used the time to speak with Colambu, who told him about the events at Limasawa. The following day, Humabon exempted the Spanish ships from taxation and even expressed interest in signing a treaty for trade and peace. The agreement was swiftly signed and followed by celebrations. Unsurprisingly, Magellan had included in the terms the establishment of a fortified trading post, which opened its doors on April 12th.[22] Just as had happened in the Indies, the locals found the prospect of buying directly from the Europeans enticing, as it bypassed the many intermediaries involved in the traditional Arab and Venetian trading routes. The iron and bronze items were particularly sought out, leading Pigafetta to write that *"they gave us ten weights of gold for fourteen pounds of iron, each weight is a ducat and a half*[23]*"*, and prompting Magellan to forbid them from accepting gold, as not to not drive its price up.[24]

Magellan's next priority was to evangelise the natives. While he was a pious man—certainly more than the average sailor of the time[25]—it is hard to believe his actions were purely driven by religious belief. Magellan was aware the Treaty of Tordesillas signed between Portugal and Spain prevented both nations to claim Christian territories, meaning that—even if his newly found Philippines was indeed on the Portuguese hemisphere—Spain could argue

---

[21] Tim Joyner, *Magellan* (Maine: International Marine, 1992), on p. 38.

[22] José Toribio Medina, *El descubrimiento del Océano Pacífico* […], on p. CCLXXIX.

[23] The Roman *libra* was equivalent to about 330 g, so according to Pigafetta the locals exchanged 15 ducats of gold (5,600 maravedies, nearly five times the monthly salary of a sailor) for less than five kilograms of iron.

[24] Antonio Pigafetta in H.E.J. Stanley, *The first voyage round the world* […], on p. 91, 92.

[25] Which often dragged their feet in fulfilling the church's commandments aboard (Pablo Emilio Pérez-Mallaína Bueno, *Spain's Men of the Sea*, Baltimore, The Johns Hopkins University Press, 1998, on p. 239).

they had already evangelised them. It was by no means a watertight argument but, then again, the Treaty was also not impermeable to interpretation.

Through a combination of elaborate ceremonies and promises of better treatment for his Christian brethren, Magellan's evangelisation efforts were remarkably effective. To set the example, Humabon was the first to be baptized by the fleet's chaplain, being given the Christian name of Don Carlos—in honour of the Spanish king. Then followed his heir (Don Fernando) and Colambu (Don Juan). Over the next eight days, the chaplain is reported to have baptised some 2,200 indigenous, including a few from neighbouring islands.[26]

However, not everyone was convinced by Magellan's trading opportunities, military power, and religion. Humabon complained that some neighbouring chieftains considered themselves his peers and would not accept his authority. Magellan sent word to the defying lords, threatening them with death and confiscation of all assets if they did not submit to Humabon and, by proxy, to Spain. Several of them refused, prompting Magellan to burn down one of their villages and confiscate their livestock.[27] A couple of them submitted after the demonstration of force but Lapulapu, the chieftain of Mactan—a small island in front of Cebu, today connected by a couple of bridges to the city—remained resolute. Worse still, he taunted Magellan to come and try to burn down his village. Humabon labelled Lapulapu as his subordinate, and Pigafetta also described him as nothing more than a chieftain. However, Humabon was married to Lapulapu's niece[28] and, while Mactan was indeed small, it was strategically positioned to be able to block water traffic to Cebu.

Whether Lapulapu was a mere chieftain or a rajah of his own right, Magellan felt compelled to accept his challenge, as backing down could jeopardise the perception of European invincibility he was trying to instil. To convince the remaining unbelievers, he decided to personally lead a small party of well-armed men to attack Mactan.[29]

This was not the first time in the expedition the captain general did something we today find abhorrent—he had, after all, abducted natives at Port San Julián and killed defenceless villagers at Guam—but it was seemingly the first time such an act was openly opposed by his officers.[30] While they did not question the righteousness of occupying territories and burning down native

---

[26] Transylvanus in H.E.J. Stanley, *The first voyage round the world* […], on p. 199.

[27] José Ibáñez Cerdá, "La Muerte de Magallanes", *Actas do II Colóquio Luso-Espanhol de História Ultramarina*, 1975, p. 410–433, on p. 428.

[28] William Henry Scott, *Barangay: sixteenth-century Philippine culture and society* (Manila: Ateneo de Manila University Press, 1994), on p. 129.

[29] Antonio Pigafetta, *The first voyage around the world* […], on p. 56.

[30] Idem.

villages, doubts were raised whether the benefits of this attack outweighed its risks. Humabon, knowing Lapulapu was a fierce adversary and that his village would be well defended, also opposed the plan. Seeing Magellan would not budge, he offered to support the Spanish landing party with his own men and war canoes, which the captain general refused. To make matters worse, the attack was poorly timed, and the three longboats arrived at the shallow bay in front of Lapulapu's village at low tide. Unable to progress through the exposed reefs, the vessels stayed at the entrance of the bay and the men trudged on foot through the water. Humabon's force was instructed to enjoy the spectacle from afar and to not intervene.[31]

The accounts of what happened next vary widely.[32] Pigafetta, who was part of the attack party, provides a first-hand detailed description of the events,[33] but the flamboyant gentleman was known for propping his stories for dramatic effect. Over the centuries, other reports of eyewitnesses have been used to cross-check Pigafetta's account and weed out the points where he likely let his imagination run free.

In a possible version of events,[34] the small but well-armed party (with around 50 men, and another 10 staying behind in the longboats) made its way to the shore unopposed. The village had been evacuated, so they started burning houses. At the sight of fire and noticing Humabon's force remained on their war canoes, Lapulapu's men came back in vast numbers to defend their village. Pigafetta mentions there were more than 1,500 natives, which seems an exaggeration for such a small island, but there is nevertheless little doubt they vastly outnumbered the attackers. Lapulapu had organised his defence force well, splitting it in several units which confronted Magellan and his men from all flanks. Despite the vast differences in weaponry—stones, stakes and poisoned arrows against harquebuses and crossbows—the Europeans could not repel the residents. After a few hours, with lead and crossbow bolts running low, Magellan gave the order for a gradual retreat. Noticing the light cannons aboard the longboats were too far out to provide covering fire, most men chose instead to run back to the boats, leaving Magellan and eight others alone on the front line. Among those who stayed behind with the captain general were Pigafetta, Enrique, and Rebelo. The latter was mortally wounded by a poisoned arrow, as the locals had realised the Europeans were

---

[31] Idem; Tim Joyner, *Magellan* [...], on p. 191, 192.

[32] José Ibáñez Cerdá provides a summary of the various accounts ("La Muerte de Magallanes" [...], on p. 413–433).

[33] Antonio Pigafetta, *The first voyage around the world* [...], on p. 56–58.

[34] Based on the views of Tim Joyner (*Magellan* [...], on p. 191–196) and Jean Denucé (*Magellan* [...], on p. 315–319), both of which have attempted to sift through the inconsistencies between contemporary accounts of what happened.

not using leg armour—left behind to make it easier to climb in and out of the longboats—and were now aiming at the exposed legs. Also wounded by a poisoned arrow and likely enraged by the death of Rebelo, Magellan pushed forward, eventually finding himself isolated and close to the enemy. He was quickly surrounded and showered with rocks. One warrior got closer and slashed the captain's leg, causing him to slump forward. The others joined and finished off Magellan with thrusts from bamboo lances. Pigafetta and Enrique made it back to the longboats, but five other crew members were not so lucky, and their bodies were left behind together with those of Magellan and Rebelo.

The captain general had thrown the dice one time too many. After surviving several mutiny attempts and a winter in Patagonia, discovering the Strait that bears his name, and leading arguably the most daring oceanic crossing of all time, Magellan died in a petty brawl.

Magellan had left a will,[35] but none of his requests were honoured. He wished to be buried on a church dedicated to Our Lady, but his body was never recovered—the survivors attempted to trade Magellan's remains for anything from the trading post, but Lapulapu refused.[36] His burial, to be accompanied by 30 masses plus food and clothing for the poor, never took place. His wife and son did not receive Magellan's profits from the expedition nor the governorship of the Philippines. They died around 1521, leaving no one to bear Magellan's name and coat of arms.[37] Rebelo, who was set to receive 30,000 maravedies from Magellan's estate, also died at Mactan. And Enrique, set to become a free man after Magellan's death, was not released from slavery.

Today, the grounds of the Battle of Mactan illustrate how the roles of hero and villain have been blurred over time. The shallow reefs which played a decisive role in the fight were named Magellan Bay, and a shrine in honour of the explorer was erected in 1866, during the Spanish colonial rule. However, in 1981, 35 years after the Philippines became an independent nation, a statue of Lapulapu was built next to Magellan's shrine. Below the warrior's statue, a plaque reads:

"Here, on 27 April 1521, Lapulapu and his men repulsed the Spanish invaders, killing their leader, Ferdinand Magellan. Thus Lapulapu became the first Filipino to have repelled European aggression."

---

[35] Francis Guillemard, *The life of Ferdinand Magellan and the first circumnavigation of the globe: 1480–1521* (New York: Dodd Mead, 1890), on p. 316–326.

[36] José Toribio Medina, *El descubrimiento del Océano Pacífico* […], on p. CCLXXXVI, CCLXXXVII.

[37] See How Did Magellan's Expedition Change the World?

12.1 What Happened to the Philippines?

> ***The short answer***
>
> *Filipinos continued to repel European aggression for a few more battles but, as their name betrays, would eventually lose the war. By the 1570s, Spain had established a permanent presence in the Philippines and stable shipping routes to and from Mexico. The Spanish Crown attempted to make Manila the epicentre of a three-pronged emporium—trading silver mined in Mexico for silk from China and spices from Southeast Asia, but on most years the colony failed to turn a profit. Still, Spain kept hold of it till 1898, when it was ceded to the United States, and the long-sought independence only came in 1946. Today, less than 1% of Filipinos speak Spanish but more than 80% are Catholic, as the evangelization effort started with Magellan at Cebu never lost thrust. So far, all attempts to change the country's name—given in honour of Philip II, Charles V's son—failed. The Philippines are still the Philippines.*

For the first decades following Magellan's expedition, Spain continued to lust over the Moluccas, treating the Philippines as a mere stop-over or a secondary objective. The first follow-up expedition, led by Loaísa in 1525, ended badly. After struggling to cross the Strait of Magellan,[38] most of the crew—including Loaísa—lost their lives to further hardships. A single ship made it to the Moluccas, where its 24 survivors eventually fell into the hands of the Portuguese. Another one wandered instead to Mindanao, where it wrecked, and the survivors were imprisoned by the locals.[39] Spain tried once more in 1526, with Sebastian Cabot, who never made it past the La Plata River.[40]

Subsequent attempts, instead of attempting to cross two oceans and a treacherous strait in one sitting, departed from Mexico and Peru. Still, making it safely across the Pacific and back would prove to be anything but easy. The first to try it was Álvaro de Saavedra, who set sail from Mexico in 1527 with a three-ship 110-men fleet, with orders to sail to the Moluccas with a stop-over at Cebu. Saavedra lost two ships to the Pacific crossing and failed to reach Cebu, but managed to reach Mindanao (where they picked three survivors from Loaísa's crew) before continuing to the Moluccas. He then tried twice to get back to Mexico but failed to find the winds to do so.

---

[38] See Did the Strait of Magellan Strait Ever Get Much Use?

[39] O.H.K. Spate, *The Spanish Lake* (Canberra: Australian National University Press, 2004), on p. 90.

[40] José Maria Moreno Madrid and Henrique Leitão, *Atravessando a porta do Pacífico* (Lisbon: By the Book, 2020), on p. 28.

Saavedra lost his life on the second attempt, and the expedition's 22 survivors returned to the Moluccas and surrendered to the Portuguese.[41]

Ten years later, in 1537, an obscure voyage with a fitting tenebrous ending left Peru, after delivering supplies and soldiers from Mexico. It is unclear what orders had been given to the captain, Hernando Grijalva. Some say he received secret instructions for another attempt to reach the Moluccas or the Philippines. Others argue he was pursuing some rumoured rich islands to the west of Peru before the crew mutinied, killed him, and headed to the Moluccas seeking personal enrichment. In any case, they would wreck in New Guinea, where the only three survivors were eventually picked up by a Portuguese ship.[42]

Besides failure, these four expeditions also had in common the fact that its survivors were rescued and repatriated by the Portuguese, often not before being roughed up in the battlefield or in a derelict prison, and serving as currency in diplomatic negotiations. Indeed, the far east was no exception to the juxtaposition of Iberian brotherhood and neighbouring rivalry that marked Spain and Portugal's relations over the course of history.

Andrés de Urdaneta, one of those survivors, would later have a decisive role in Spain's colonisation of the Philippines. Urdaneta was part of Loaísa's 1527 fleet and lingered in the Moluccas till 1535, when the Portuguese agreed to repatriate the surviving Spaniards. Urdaneta boarded a Chinese junk bound for Java, and then travelled northwest first to Malacca and then to Cochin, where a Portuguese ship took him to Lisbon. At the advice of the Spanish ambassador he quickly left Portugal, but not before having his maps and journals confiscated by the local authorities. Still, nobody could erase what he had seen, and once back to Spain he enthusiastically described not only the riches of the Moluccas, but also the untapped potential of the Philippines.[43]

After the string of failures, Urdaneta's testimony helped to rekindle the interest in the Philippines or, as the Spaniards called them at the time, *Islas del Poniente* (Islands of the West[44]). In 1542, another expedition departed from Mexico, this time specifically targeting these islands, rather than the Moluccas as before. Ruy López de Villalobos, the captain in charge of a fleet of six ships and over 370 men, was given three objectives: establish a hub for trading with China and Japan, evangelise the natives, and find a viable return route. Villalobos had set his sights on Cebu but ignored his pilots'

---

[41] O.H.K. Spate, *The Spanish Lake* […], on p. 90.

[42] Idem, on p. 96, 97.

[43] Mitchell Mairin, *Friar Andres de Urdaneta, O.S.A.* (London: Macdonald and Evans, 1964), on p. 13, 63–73.

[44] That is, west of their American colonies.

advice and ended up instead at Mindanao, well outside the trading routes and with no favourable winds to sail north to Cebu. The locals, for loyalty or for fear of the Portuguese, were not inclined to help the Spaniards with directions and supplies. Eventually Villalobos and his party set up a precarious base in Sarangani, on the south of Mindanao, but food was scarce. After resorting to eating anything they could get their hands on, a menu apparently including poisonous lizards which claimed the lives of several mariners, Villalobos ordered one of his ships to go back to Mexico and return with supplies. The attempted failed, and so did a second one. At the end if their wits, the Spaniards saw no other option than to make their way down to the Moluccas, where they once more fell into the hands of the Portuguese.[45]

At an enormous cost in lives and resources, each one of these failed voyages brought back insights on where to establish a base (Cebu or one of the nearby islands, further away from Portuguese influence and with ample supplies) and on how region's winds and currents changed throughout the year (essential for a successful return trip to the Americas). By then, the Spaniards had also settled on the archipelago's current name. Villalobos or Bernardo de La Torre (one of his captains) christened Leyte, Samar and the surrounding islands as *Filipinas* (Philippines) in honour of Charles V's son,[46] and the name was eventually extended to the entire archipelago.

It took more than 20 years for the next expedition to leave Mexico. It was 1565, and Charles V had abdicated the throne in favour of his son, Philippe II, who selected Miguel López de Legazpi to lead four ships and 380 men. Legazpi had not been the king's first choice, as that honour had been bestowed on Urdaneta, by now an Augustinian monk and well versed on navigation matters. Urdaneta declined the responsibility but accepted to join the expedition as an advisor. Once 100 leagues out into the ocean, their sealed orders were opened, spreading consternation through the crew. In place of the rumoured voyage to China, the ships were to set sail towards the Philippines, a destination which had so far not been kind to the Spaniards. They were tasked with identifying locally grown spices and other trading opportunities, and then finding a return way to Mexico. If possible, they should also establish a settlement and evangelise the natives.[47]

---

[45] O.H.K. Spate, *The Spanish Lake* […], on p. 97, 98.

[46] Villalobos also named Mindanao, which they reached first and assumed to be one of the archipelago's main islands, as *Caesarea Karoli*, in honour of Charles V, as he was not only Spain's king but also the Holy Roman Emperor (O.H.K. Spate, *The Spanish Lake* […], on p. 98).

[47] O.H.K. Spate, *The Spanish Lake* […], on p. 100–102.

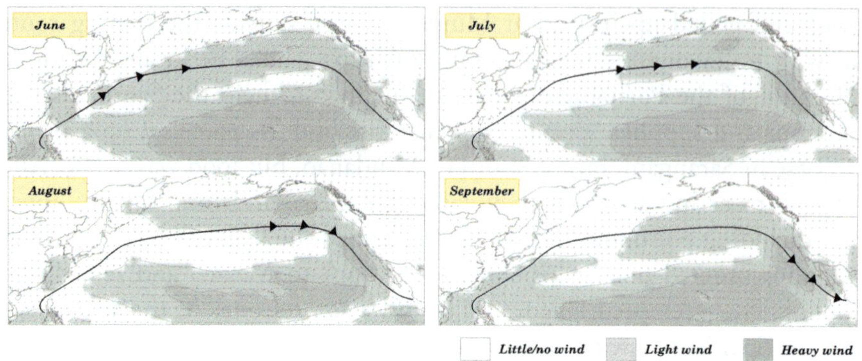

Fig. 12.2 Urdaneta's route back to the Americas overlaid with the prevailing winds for each month of the voyage, illustrating how good timing is key for a successful passage[50]

The mission was not off to the best start as, shortly after the controversial instructions were shared, the *San Lucas,* the fleet's smallest ship, mysteriously disappeared. The remaining three ships soldered on and were able to reach Cebu, but the residents were far from welcoming, as the Portuguese had been conducting raids in the region masquerading as Spaniards. After a brief bombardment though, the town was occupied and the locals watched hopelessly as the newcomers broke ground for a fort.[48]

Nearly 45 years after Magellan's expedition, Spain was back to Cebu. However, the regained foothold would be of little use as a trading hub with China if they could not find a viable return pathway to Mexico. On June 1st of 1565, after a mere three weeks of preparation, Urdaneta boarded the *San Pedro* in hopes of succeeding where others had not. After carefully navigating the treacherous waters between the Philippine islands, Urdaneta reached the open sea and made his way up north. The time window was just right to catch the favourable monsoon westerly winds but avoid the typhoons (Fig. 12.2), and the *San Pedro* managed to reach the Americas after more than 130 days, during which scurvy killed or incapacitated a good part of the sailors.[49]

The arrival to Mexico was not as triumphal as Urdaneta expected. The *San Lucas*, the ship that had vanished at the start of the expedition, was already there and claiming to have made it all the way to the Philippines and back. Suspecting desertion, the Spanish authorities opened a formal inquiry but

---

[48] Idem, on p. 102.

[49] Mitchell Mairin, *Friar Andres de Urdaneta, O.S.A* […], on p. 137, 138.

[50] Wind speeds and direction averaged for the 1991 to 2020 period, for each month ("IRI/LDEO Climate Data Library", International Research Institute for Climate and Society, 2024, https://iridl. ldeo.columbia.edu/).

could not reach a definitive conclusion. Alonso de Arellano and Lope Martín, respectively the captain and the pilot of the *San Lucas*, provided details of the geography of the Philippines, suggesting they had indeed been there. However, other elements of their accounts seemed farfetched, including the claim they had wandered around the archipelago for months without anyone seeing them.[51] The incident casted doubts on which one of the ships had inaugurated the return route but took little away from the wider picture: now Spain had a viable return route to and from the Philippines.

Meanwhile, back at Cebu, Legazpi had succeeded in establishing more amicable terms with the natives and was busy resuming Magellan's strategy of imposing Spanish lordship by means of the sword and the cross. The approach was perilous, as Lapulapu had vividly illustrated 45 years ago, but could be quite effective. By exploiting existing feuds between local chieftains, the Spaniards, despite being vastly outnumbered, were able to command vassalage from adjacent islands, first by military control and then by religious rule. Some proved to be more tenacious than others, particularly those in the southern islands of Mindanao and Sulu, and the Portuguese in the Moluccas were also not happy in having their Iberian neighbours so close by. Eventually, tired of these hurdles and by the limited supplies of food, Legazpi moved from Cebu to Panay. In 1570, he continued his northern migration and finally settled in Manila, which would become the capital of the Spanish East Indies.[52]

Spain finally had a foothold in the region. It had taken 50 drawn-out years though, and Lapulapu likely was not around anymore to witness the triumph of his foes. They would eventually colonise the entirety of modern-day Philippines, but some territories were harder to control. Parts of Mindanao and Jolo, for instance, only came under the effective control of the empire in the last quarter of the nineteenth century, nearly 300 years after the arrival of Legazpi.[53] Several islands dotted across the Pacific which Magellan or subsequent explorers found were also annexed to the Spanish East Indies, including Guam, the Northern Mariana Islands, Palau, and the Marshall Islands. Attempts to expand into other directions were less successful. To the south, the take-over of North Sulawesi (today part of Indonesia) failed. To the north, portions of Taiwan were only briefly in the fold. And to the west, the attempt to gain a foothold in mainland Asia went no further than a brief presence in Cambodia.

---

51 O.H.K. Spate, *The Spanish Lake* [...], on p. 104, 105.

52 O.H.K. Spate, *The Spanish Lake* [...], on p. 102, 103.

53 United States War Department, *Annual Report of the Secretary of War* (Washington, D.C.: U.S. Government Printing Office, 1903), on Volume III, p. 379–398.

Spain's opposition in Southeast Asia was not limited to local kingdoms and the Portuguese. The presence of the Dutch, who would become the main colonising power in the region, was unintendedly accelerated by the Spaniards themselves. In the second half of the sixteenth century, the Dutch rebelled against the rule of Philipp II, who accumulated the lordship of the Netherlands with the Spanish throne. To sever the trading which financed the uprising, Philip II closed the Iberian ports to Dutch ships, but the strategy backfired. Up to that point, the Dutch's commercial ambitions had been limited to intermediation, buying spices and other eastern products in the Iberian ports and shipping them to Northern Europe. With the route blocked, they instead took advantage of their maritime experience and first-hand knowledge of the Indies—gained through privateering and Dutch mariners in Iberian expeditions—to go directly to the source.[54]

Despite the constant attacks from local kingdoms and the Dutch, Spain held on dearly and hoped to control three of the most important sixteenth century global commodities: silver, silk, and spices. Silver from American mines was shipped from Acapulco to Manila, where it was used to buy silk from China and spices from Southeast Asia. These commodities were then sent back across the Pacific to Acapulco, transported over land to the west coast, and crossed over the Atlantic to Europe, where demand from them soared. A flurry of other sought out products could also be included in the convoy, such as gems, iron, ceramics, wax, palm wine, tea, hemp, or cotton. It seemed that Mexico, which sat at the middle of these two oceanic routes and held administrative control over the Spanish East Indies, would become the new epicentre of the world.[55]

That never happened, though. Manila did not rise to become *the* emporium of the east, but rather one of several, facing stiff competition from Malacca (first controlled by the Portuguese and then by the Dutch) and Macao (established by the Portuguese in 1557), as well as from local trading routes, which were shaken but never broken by the arrival of the Europeans. The promise of benefiting from local products also fell short, as the Philippines and the other islands which made up the Spanish East Indies had scarce natural resources.[56] Except for three decades in the eighteenth century, these territories were always a drain on the royal treasury, requiring annual subsidies

---

[54] Peter Emmer, "The First Global War: The Dutch versus Iberia in Asia, Africa and the New World, 1590–1609", *e-JPH*, 1(1), 2003, on p. 6, 7.

[55] Herb Kawainui Kane, "The Manila Galleons", in Bob Dye (ed.). *Hawai'i Chronicles: Island History from the Pages of Honolulu Magazine.* Volume I (Honolulu: University of Hawaii Press), on p. 25–32.

[56] O.H.K. Spate, *The Spanish Lake* […], on p. 104.

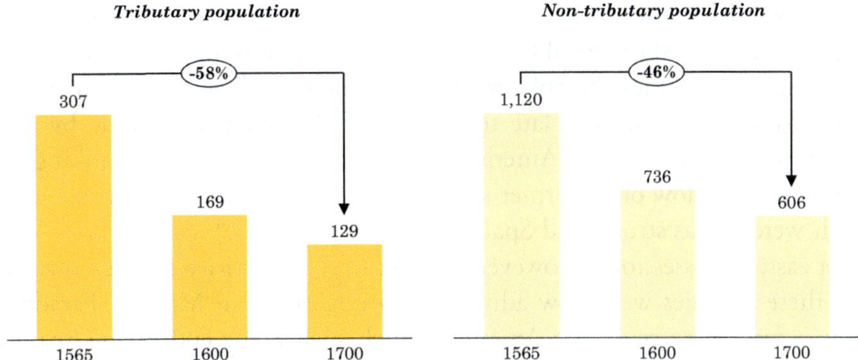

Fig. 12.3 Comparison of the 1565 to 1700 population decline in the Philippines for both tributary (i.e., under the *encomienda* regime) and non-tributary regions[61]

to fund the military and the clergy maintaining territorial integrity.[57] Abandoning or selling the Philippines was routinely discussed but never taken too far, as this would crumble the financial interests of a local elite of ecclesiastics, military, and merchants cornering the Manila-Acapulco trading, and of *encomenderos* with the right to exact tribute from the natives.[58]

Slavery had been formally abolished in the Spanish Empire before the occupation of the Philippines had even begun, but it quickly became apparent the survival of the colony required some alternative way to benefit from the labour of the native population. The solution was to introduce the *encomienda* system already in use in the Americas, which granted selected individuals the right to collect tax from the native population. In turn, these *encomenderos* provided protection and education,[59] an exchange which rarely swinged in favour of the locals. Between 1565 and 1700, the tributary population (i.e., those under the *encomienda* regime) of the Luzon and Visayas archipelagos dropped by nearly 60%, from ~310,000 to ~130,000, compared to 45% in non-tributary areas (Fig. 12.3). Much like what happened in the Americas,[60] bloodshed from the conquest and Old World diseases also had a profound impact on the native population, reducing it from an estimated ~1.4 m pre-1565 to ~0.7 m by 1700.

---

[57] Idem, on p. 157.

[58] Linda Newson, *Conquest and pestilence in the early Spanish Philippines* (Honolulu: University of Hawai 'i Press, 2009), on p. 8.

[59] Idem, on p. 7.

[60] See What Happened to the Natives Encountered by the Fleet?

[61] Linda Newson, *Conquest and pestilence in the early Spanish Philippines* [...], on p. 255, Table 14.1.

The Spanish Empire began to shake in the early nineteenth century when Napoleon abducted the royal family and the liberal movement gained traction across Spain's colonies. Napoleon's defeat brought back absolutism to mainland Spain, but it was too late to stop the overseas tsunami. One by one, the Spanish colonies in the Americas fought for independence, shrinking the empire to a shadow of its former self (Fig. 12.4). In the Pacific, the echoes of revolt were not as strong and Spain still held on to the Philippines and to its other eastern possessions. However, the link to the Americas has been severed, and these colonies were now administered directly from Madrid. Paradoxically, it was an event in the Americas which spelled the end of the Spanish East Indies. In 1898, rebels threatened to overthrow the Spanish government from Cuba, which was, together with Puerto Rico, its only remaining American colony. USS Maine, a battleship sent to Havana to protect the United States interests in the region, exploded and sank, killing three-quarters of its crew. It was unclear if the explosion had been an accident or an act of sabotage,[62] but the uncertainty was enough to engulf Spain and the United States in a full-blown conflict. The war between the declining European power and the rising American nation was swift and ended with a decisive victory for the latter. The capitulation terms included the sale of the Philippines, together with Puerto Rico and Guam. The reminder of the Spanish East Indies, an assortment of over 6,000 scattered islands which included the Northern Marianas, Palau and the Caroline Islands, were sold to Germany.

After more than 300 years, the Spanish rule initiated by Legazpi had come to an end. But independence was not yet within the grasp of the Filipinos. The First Philippine Republic, established shortly after the end of the Spanish-American War, was not recognised by the United States, prompting the Philippine-American War. The conflict ended in 1902, with a commitment from the occupying force that they would eventually leave the territory. The occupation by Japan during World War II delayed independence, which happened only in 1946.

Today, less than 1% of Filipinos speak Spanish,[63] and Filipino (a standardised version of Tagalog, a native language) and English are instead the country's official languages. This is remarkably different from what happened in most other former Spanish colonies, who have kept to the Spanish language. Yet, it would be rash to assume the 300 years of occupation left no

---

[62] The cause of the explosion remains unclear to this day, with multiple analyses (ranging from official enquiries in the nineteenth and twentieth centuries to a National Geographic investigation in 2002) often arriving to contradictory results.

[63] Clarissa Andrés Barrenechea, "La enseñanza del español como lengua extranjera en Filipinas. Estudio de caso de la Universidad Ateneo de Manila", *Revista Filipina*, 1(1), 2013/2014, on p. 37.

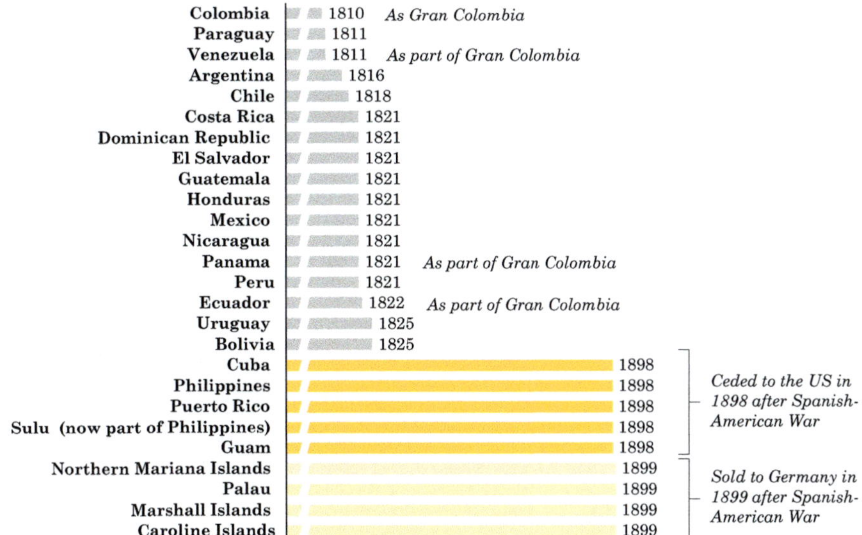

Fig. 12.4 Dates when Spanish colonies declared independence or were ceded

trace. Apart from the hints of Spanish heritage found everywhere, from food to surnames, religion is by far the most obvious European inheritance. The evangelisation efforts initiated by Magellan at Cebu and resumed by Legazpi in 1565 were tremendously effective, as around 80% of the country is now Catholic,[64] one of the highest percentages in the world.

Filipinos show no inclination to forego their religion, but other elements of Spanish heritage are embraced with less fervour. Their very name, a direct reference to Phillip II, is perhaps the biggest thorn in the republic's identity. Changing the name of a country is uncommon but not unheard of, as the likes of Sri Lanka (previously Ceylon) and Cambodia (old Khmer Republic) have done it. All renaming attempts so far have failed though, usually due to the difficulty in finding a common identity, since before the occupation the archipelago was not a single nation but a medley of different tribes and peoples. In the 1960s a Senate bill proposed 'Malaysia',[65] a reference to the region's Malay ancestry. But the Federation of Malaysia used the name first, and today Malaysia is Malaysia, and the Philippines continues to be the Philippines.

---

[64] "Religious Affiliation in the Philippines (2020 Census of Population and Housing)", Philippine Statistics Authority, 2020, https://psa.gov.ph/content/religious-affiliation-philippines-2020-census-population-and-housing.

[65] Minako Sakai, "Reviving Malay Connections in Southeast Asia" in Elizabeth Caol (ed.), *Regional Minorities and Development in Asia*, (London: Routledge, 2009), on p. 124.

13

The Ambush
Cebu, Philippines (April 27th, 1521 to May 1st, 1521)

Magellan's death shattered the notion of Spanish invincibility, and likely brought to the surface brewing issues. As had happened at Rio de Janeiro, the crew chased down native women at Cebu (Fig. 13.1). Unlike at Rio though, Magellan's efforts to get his crew to behave—which included punishing his relative Duarte Barbosa and removing him from the *Victoria's* command— had been largely unsuccessful, prompting the enraged natives to complain to Humabon.[1] After Magellan's death, the officers' decision to elect the punished Barbosa to replace the fallen leader (sharing the command with Juan Serrano)[2] poured gasoline into the fire.

Any trace of respect Humabon still had for the Spaniards disappeared when Barbosa and Serrano decided to close the trading post and set sail, leaving Humabon alone to fend off his enemies.[3] Enrique also experienced first-hand the crassness of the new leaders. While he was recovering from the wounds sustained fighting alongside Magellan, Barbosa ordered him to go ashore and aid the negotiations for two pilots and provisions. When Enrique declined, on account of his wounds, an enraged Barbosa threatened the slave with flogging. Worse still, he told Enrique that, rather than freeing him as per Magellan's will, he would personally see he was returned to the widow as soon as they were back to Seville. Pigafetta states it was in this moment Enrique decided to conspire with Humanon to betray the Spaniards and seize their ships and

---

[1] Jean Denucé, *Magellan: la question des Moluques et la première circumnavigation du globe* (Brussels: Hayez, 1911), on p. 324.

[2] Jean Denucé, *Magellan* [...], on p. 323.

[3] Idem.

R. Gaspar, *The Revolution of Magellan*, https://doi.org/10.1007/978-3-032-10797-8_13

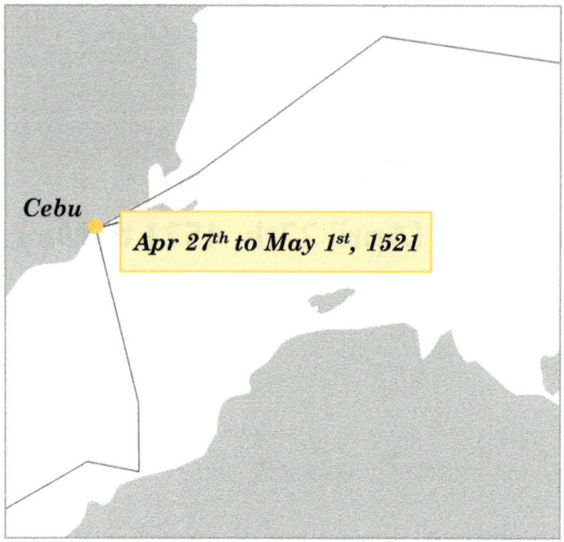

Fig. 13.1 Fleet route[4]

cargo.[5] He had no way of knowing what was going inside Enrique's head but, judging by what happened next, he may have been close to the mark.

On May 1st, Enrique returned to the *Trinidad* bringing an invitation from Humanon. To honour their lifetime alliance, the rajah wanted to gift a collection of exquisite jewellery to the Spanish king, to be trusted to the high command over a ceremonial banquet. Serrano was suspicious but, after seeing his hesitations mocked by Barbosa, agreed to go. More than 20 others, both officers and sailors, also hopped onto the longboat and made their way to the shore. A notable absence was Pigafetta—not one to miss a party, he had been forced to stay behind nursing his wounds from Mactan.[6]

Magellan, if still alive, would have likely remembered his plights in 1509 and shared Serrano's suspicions. The fallen captain general and his friend Serrão had been part of Portugal's first attempt to take control of Malacca. The same tactic likely at play at Cebu—signing a peace treaty and establishing a trading post, with the goal of returning later with war ships—seemed to be going well, but the sultan had been warned by Arab merchants not to

---

[4] Fleet route adapted from Tomás Mazón Serrano ("Mapas", Ruta Elcano, June 19, 2024, https://rut aelcano.com/mapas/); map adapted from Natural Earth (1:10 m large scale land data, version 5.1.1, https://www.naturalearthdata.com/downloads/).

[5] Antonio Pigafetta, *The first voyage around the world 1519–1522*, edited by Theodore Cachey Jr. (Toronto: University of Toronto Press, 2007), on p. 59.

[6] José Toribio Medina, *El descubrimiento del Océano Pacífico: Vasco Núñez de Balboa, Fernando de Magallanes y sus compañeros* (Santiago de Chile: Imprenta Universitaria, 1920), on p. CCC.

trust the Portuguese. The apparently obliging monarch told Diogo Lopes de Sequeira, the Portuguese commander, he so happened to have a full cargo of pepper waiting in a warehouse. If the commander's crew could help transport it to the port, they could immediately do business. Sequeira, despite being advised caution by Chinese merchants, did not judge the opportunity too good to be true and merrily agreed to the trade. With a sizeable landing party already in the warehouse, Garcia de Sousa—one of Sequeira's captains—became increasingly suspicious of the large number of Malay boats swarming around the Portuguese ships. Magellan was sent to the flagship to warn Sequeira, and then quickly returned to the shore to aid his comrades. Part of the landing party had already been trapped in the warehouse by the sultan's men, but some managed to fight their way back to the beach, where Magellan helped them board a longboat. Serrão was among the men Magellan saved, but more than half of the landing party was killed or captured.[7]

Back at Cebu, the backdrop for a similar drama was taking shape. As the landing party was being ushered to the ceremonial house, two of the officers—Carvalho and Espinosa—suspected foul play and returned to the ships to get reinforcements. While doing so, they heard shouts and cries from the shore. Carvalho, now the most senior officer, ordered the gunners to attack the town with the ship's cannons. Serrano, tied down and bleeding, was dragged outside to stop the shelling. Humanon's men had sliced the throats of everyone except Enrique, he said, but were willing to exchange his life for a pair of pieces of artillery. Carvalho complied, but Humanon raised the stakes after receiving the ransom. As they were negotiating a further pay out, Serrano reportedly warned his colleagues the natives were stalling until reinforcements came to seize the ships. Heeding the warning, Carvalho gave the order to set sail.[8]

More than 20 men were left behind. Contrary to Serrano's statement that everyone except Enrique had been slain, a survivor of the later 1525 Loaísa expedition reported up to eight survivors being sold as slaves in China.[9] Whether slain or enslaved though, they were lost to the expedition. The remaining crew would now have to navigate through unfamiliar waters without their three captains (Barbosa, Serrano, and Góis), without San Martín (the fleet's astronomer and the only one capable of making

---

[7] Fernão Lopes de Castanheda, *História do descobrimento & conquista da Índia pelos portugueses* (Coimbra: João Barreira and João Álvares, 1554), on Livro II, Capítulo CXVI, p. 221–225; João de Barros, *Década segunda da Ásia de João de Barros* (Lisbon: Jorge Rodriguez, 1628), on Livro IV, Capítulo IV, on f.93–96.

[8] Tim Joyner, *Magellan* (Maine: International Marine, 1992), on p. 197, 198.

[9] Emma Blair and James Robertson, *The Philippine Islands*, Volume 2, Number XXXIV (Cleveland: A.H. Clark Company, 1903), on p. 461.

longitude measurements) and without Enrique (the only one who spoke a local language). From the original assignments, Carvalho was the only pilot left[10]: Estevão Gomes had deserted with the *San Antonio*, Vasco Galego and Juan Rodríguez de Mafra had died from scurvy, and Serrano—originally performing double duty as captain and pilot of the small *Santiago*, was left on the distancing Cebu shore.

13.1   What Drove Men to a Life at Sea?

> **The short answer**
>
> *It was no secret there was a myriad different ways to die abord a sixteenth century ship, and Magellan struggled to put a crew together. Eventually he managed to do so, prompting a question: why were these men willing to die in a remote part of the world? Compensation, which included a steady salary and a share of the sale of cloves, certainly played a role. Still, the money fell short of the risks, particularly for the low-ranking mariners. Instead, most of these were driven by despair, having exhausted other ways of earning a living on land.*

Some 265 men left Spain aboard the confined living spaces of the five ships. The exact crew size is not entirely clear, as not even the fastidious book-keeping procedures of the *Casa* were able to keep up with the daily barrage of desertions, dismissals, and enlistments.[11] It was not easy convincing men to join a multi-year voyage with mysterious objectives, but eventually the deed was accomplished.

Little more than 90 of this original crew of 265 would survive,[12] a death rate of 66%—close to the 68% of the Titanic.[13] Applicants had of course no way of foreseeing it, but it was a poorly kept secret that sixteenth century maritime exploration was a risky endeavour. It would remain so for quite a long time: for instance, of the 1,033 recorded Portuguese ships that which the round trip from Lisbon to the Indies between 1497 and 1650, over 20%

---

[10] Plus Albo, an acting pilot initially hired as a master's mate.

[11] This estimate is from Navarrete, based on the *Casa's* fleet rosters (around 240 men) and adding last-minute additions from other sources (Martín Fernández de Navarrete, *Coleccion de los viajes y descubrimientos que hicieron por mar los españoles desde fines del siglo XV*, Madrid, Imprenta Nacional, 1837, in Tomo IV, p. 12–22).

[12] Guadalupe Fernández Morente, "Los hombres de la primera vuelta ao mundo", *Carlos y el mar: el viaje de circunnavegación de Magallanes-Elcano y la era de las especias*, 2021, on p. 142.

[13] Lord Mersey, *The Loss of the Titanic, 1912* (London: The Stationary Office, 1999), on p. 110, 111.

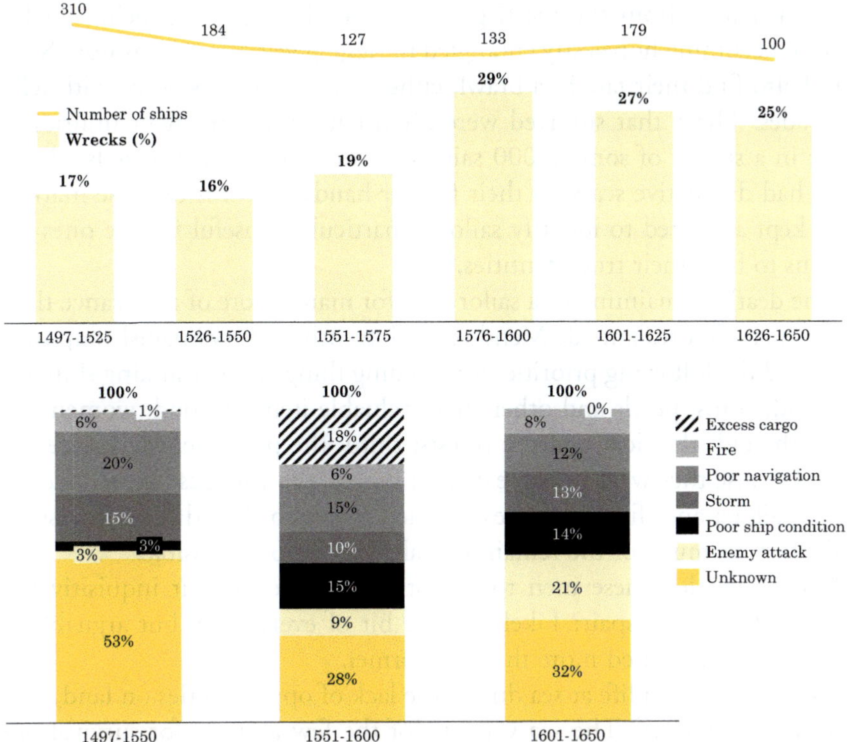

Fig. 13.2 Number of wrecks and underlying causes for Portuguese ships sailing to and from the Indies between 1497 and 1650[14]

sank somewhere along the route (Fig. 13.2). One would assume the rate of shipwrecks decreased over time as the route became more familiar, but the opposite happened due to a myriad a factors, from overloaded ships to enemy attacks. But perhaps the direst aspect of these numbers is that the cause for almost 40% of these wrecks is unknown. In most cases, these were ships that simply vanished without a trace, leaving no survivors to tell what had happened.

Even in ships making it to good port there was no shortage of deadly occupational hazards. Disease and malnutrition were the most common ones, followed by falling overboard—even if somebody noticed, the lack of manoeuvrability of the ships made it hard to quickly turn back to rescue the drowning sailor, many of which did not know how to swim.[15] During

[14] Paulo Guinote, "Ascenção e declínio da Carreira da Índia", *Vasco da Gama e a Índia*, edited by José Garcia e Maria Helena Pinto (Paris: Fundação Calouste Gulbenkian, 1999), on Volume II, p. 7–39.

[15] Richard Mandell, *Sport: A Cultural History* (New York: Columbia University Press, 1984), on p. 179, 180.

a storm, falling from the mast, getting struck by lightning, being hit by a spar, or becoming hopelessly entangled in rigging was not uncommon. Sailors could also find their fate in a brawl, either with enemy vessels or with fellow comrades. Those that survived were often left with permanent injuries and scars: in a survey of some 2,000 sailors working on Spanish vessels, close to 50% had distinctive scars on their face or hands. Records of these markings were kept and used to identify sailors—particularly useful for the ones with reasons to hide their true identities.[16]

The death or maiming of a sailor was, for many, more of a nuisance than a worry. Juan de Escalante de Mendoza, a sixteenth century Spanish shipowner, proposed the following priorities for rescuing things from a sinking ship: first, the gold, coins, pearls and other small valuable items; second, women, children, the old, the sick, and the priests; third, the passengers and slaves (the latter because they were valuable cargo); fourth, the youngest apprentices and oldest sailors; and, finally, the rest of the crew, who should nevertheless take with them as much of the remaining valuable cargo as possible.[17]

What then led these men to accept such a life? Was it inquisitiveness? Bravery? Greed? Despair? Likely a little bit of everything, but arguably the latter reasons weighted more than the former.

Most turned to a life at sea due to the lack of opportunities on land, many from a young age.[18] The sea was one of the few avenues for survival open to the masses of abandoned and orphaned children who roamed aimlessly through Seville. Some would also become apprentices involuntarily, sold by their parents or kidnapped by unscrupulous masters and ship lords. Misery was not a sailor's only mistress. Among those seeking a life at sea, there was also no shortage of men who were running away from justice or debts, and misfits and rebels would often find their way into an oceanic ship.[19]

In the late sixteenth century, well after Magellan's expedition, the appeal of a career at sea had not improved much. Diego García de Palacio, a Spanish author of naval construction and seafaring treatises, attempted to rally interest for the sea by first admitting *"death is always three or four fingers away, which is the width of the ship's planking"* but then noting that *"the worst harm it*

---

[16] Pablo Emilio Pérez-Mallaína Bueno, *Spain's Men of the Sea* (Baltimore: The Johns Hopkins University Press, 1998), on p. 74, 255.

[17] Juan de Escalante de Mendoza, *Itinerario de navegación de los mares y tierras occidentales 1575* (Madrid: Museo Naval, 1985), on p. 280.

[18] See What Was the Life of a 16th Century Sailor Like?

[19] Pablo Emilio Pérez-Mallaína Bueno, *Spain's Men of the Sea* […], on p. 23–34.

brings lasts only until death".[20] To no one's surprise, few found Palacio's words inspiring.

The noblemen who wandered onto a ship did so for similar reasons. There were of course no sons of aristocrats kidnapped to serve on board, but there were plenty of nobles—particularly of low rank or with little wealth behind their title—seeking money and the king's good graces. Serrano, abandoned in Cebu after the ambush, left behind a wife and daughters in poverty.[21]

Magellan started recruiting almost one year before the expedition left Seville and could have used more time. As if finding able seamen for an obscure two-year voyage was not hard enough, the captain general also had to wrestle crew away from two other missions: Andrés Niño's fleet heading to the Americas, and a squadron leaving for Levant from nearby Málaga.[22] It was an uneven competition, as most recruits preferred to serve under a Spanish captain than to follow a Portuguese one to uncharted waters. In any case, pickings were not much to begin with, as Castille did not have a strong maritime tradition and was still recovering from the devastating effects of the fourteenth century Black Plague.

Faced with the challenge of hiring Spaniards, Magellan turned to other nationalities, Portuguese and Italians in particular. The *Casa* did not complain too much about the latter but they did object to the former, fearful a Portuguese captain with a large entourage of his own countrymen could be tempted to defend rival interests. The Spanish king agreed, imposing a quota of five Portuguese for the fleet.[23]

While Magellan did remove some Portuguese from the rosters—including two pilots—he mostly wiggled his way around the royal decree, arguing there were simply not enough Spanish mariners available. He was not alone in this plight, as throughout the rest of the sixteenth century many Spanish captains would be forced to fill their ships with a significant proportion of Portuguese and other foreigners. Magellan's final roster had close to 45% of foreigners, including some 30 Portuguese—perhaps more, as it is likely that some disguised their true origins to circumnavigate the imposed restrictions. Among the 165 Spaniards, there were around 30 Basques, most of which did not speak much Castilian.[24]

---

[20] Diego García de Palacio, *Instrucción náutica para navegar* (Madrid: Ediciones Cultura Hispanica, 1944), on f.2.

[21] Pablo Emilio Pérez-Mallaína Bueno, *Spain's Men of the Sea* […], on p. 19.

[22] José Toribio Medina, *El descubrimiento del Océano Pacífico: Vasco Núñez de Balboa, Fernando de Magallanes y sus compañeros* (Santiago de Chile: Imprenta Universitaria, 1920), on p. CLII.

[23] José Toribio Medina, *El descubrimiento del Océano Pacífico* […], on p. CLIV.

[24] Tim Joyner, *Magellan* (Maine: International Marine, 1992), on p. 104, 252–263.

With second and third pickings aplenty, an elite team it was not. Again, Magellan was not the only captain with this problem, prompting Antonio de Guevara—King Charles V's chronicler—to cry out, a couple of decades later, that "*the sea is the cloak of sinners and the refuge of malefactors*".[25]

It was clear a life at the sea was not the first thing springing to the mind of ambitious young men, but perhaps the wages were higher than on land, somewhat making up for the extra risks and hardships?

A sailor in Magellan's expedition earned 1,200 maravedies per month (Fig. 13.3), or 40 maravedies per day. He was also entitled to three daily meals, valued at 11 maravedies.[26] The resulting daily wage of 51 maravedies was higher than the 35–45 maravedies sailors in the Spanish armadas earned in early sixteenth century,[27] but considerably less than the 85 maravedies carpenters, calkers and masons received in Seville.[28] The comparison was slightly more enticing for apprentice sailors, who earned 38 maravedies per day aboard Magellan's ships,[29] rather than the 34 maravedies paid to unskilled labourers in Seville.[30]

The other tactic used to make the risky endeavour more palatable was to offer *quintaladas*, which was cargo space the crew could use to bring back goods for trading. In Magellan's expedition, a sailor was entitled to 2 *quintales* of cloves, worth approximately 38,100 maravedies back in Seville,[31] almost as much as the sailor would earn in base salary on a three-year expedition. The *quintalada* increased with rank—from 0.75 *quintales* for cabin boys to 80 *quintales* for the captain general (Fig. 13.3).

Magellan, if he had survived, would have earned some 1,455,000 maravedies (more than the cost of acquiring the fleet's five ships second hand[32]), plus a share of the profits of newly discovered territories.[33] It was a mouth-watering amount of money, one which certainly played a role in

---

[25] Antonio de Guevara in José Luis Martínez, *Pasajeros a Indias: Viajes transatlánticos en el siglo XVI* (Mexico City: Fondo de Cultura Económica, 1983), on p. 229.

[26] Earl Hamilton, "Wages and Subsistence on Spanish Treasure Ships, 1503–1660", *Journal of Political Economy*, Vol. 37(4), 1929, p. 430–450, on p. 446.

[27] Pablo Emilio Pérez-Mallaína Bueno, *Spain's Men of the Sea* […], on p. 115.

[28] Earl Hamilton, "Wages and Subsistence on Spanish Treasure Ships, 1503–1660" […], on p. 443, Table II.

[29] 800 maravedies per month, or 27 per day, plus the ratios valued at 11 maravedies.

[30] Earl Hamilton, "Wages and Subsistence on Spanish Treasure Ships, 1503–1660" […], on p. 443, Table II.

[31] A *quintal* of cloves (approximately 46 kg) was valued at 12,700 maravedies (net of taxes and alms), or 38,100 for the 3 *quintales* alloted to sailors (Pablo Emilio Pérez-Mallaína Bueno, *Spain's Men of the Sea* […], on p. 101).

[32] See How Advanced Were Magellan's Ships?

[33] Between 4 and 7%, see Worse News.

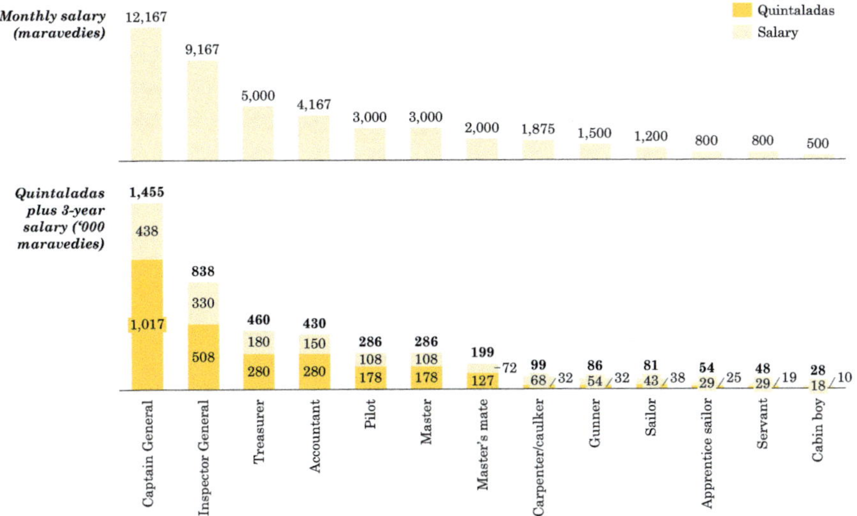

Fig. 13.3   Top: monthly salaries in Magellan's expedition; Bottom: total pay for a 3-year voyage, including both the salary and *quintaladas* (share of the sale of cloves)[34]

driving the captain general forward at any cost. Compensation fell precipitously with rank, though, down to the cabin boys who earned 50 times less than Magellan. At a 2-in-3 chance of dying, it was a rotten gamble.

---

[34] Adapted from: Tim Joyner, *Magellan* […], on p. 252–263; Cristóbal Bernal, *Crónicas de la Primera Vuelta al Mundo, según sus Protagonistas*, 2016, on p. 168, 169 (transliterated from "Informaciones sobre sueldos, mercancías y mercedes relativas a la Armada a la Especiería organizada por Fernando de Magallanes", Seville, Archivo General de Indias, ES.41091.AGI//CONTADURIA,425,N.1,R.1, c.1524, https://pares.mcu.es/ParesBusquedas20/catalogo/description/7344884).

14

The Moluccas

From Cebu, Philippines (May 1st, 1521) to the Moluccas (November 8th, 1521)

With the death of Magellan, the priorities of the expedition changed. Unlike the fallen captain general, the survivors did not benefit directly from the evangelisation and occupation of the Philippines. Perhaps the Spanish captains, if still alive, would be sufficiently close to the king to care about it. The surviving crew, however, was now likely focused on the *quintaladas*, their share of profits from the sale of cloves.

The chart Magellan had shown the king showed the location of the Moluccas (see Fig. 1.3, Chap. 1). The quickest way to get there would be to travel south until they reached the equator and then—given the challenges of determining longitude—search left and right along the equator until they found the islands.

Instead, the officers decided to go westward, a decision that remains unclear to this day (Fig. 14.1). Perhaps they no longer trusted the charts, either because they suspected Magellan withheld the true location of the Moluccas to ensure he remained indispensable, or because they had received contradicting intelligence from the locals. Whatever the reason, they seemed to have their sights now set on the Island of Borneo, which they knew from intelligence collected at Cebu to be one of the region's major trading ports,[1] and as such a good place to repair the ships and pinpoint the location of the Moluccas.

Before they could do that though, there were pressing matters to attend to. The crew was now so reduced it could not safely man three ships. Off

---

[1] Jean Denucé, *Magellan: la question des Moluques et la première circumnavigation du globe* (Brussels: Hayez, 1911), on p. 327.

R. Gaspar, *The Revolution of Magellan*, https://doi.org/10.1007/978-3-032-10797-8_14

Fig. 14.1 Fleet route[2]

the coast of Bohol Island, they decided to get rid of the *Concepcion*, which was in worse shape. After its crew and cargo were redistributed across the *Trinidad* and the *Victoria*, the *Concepcion* was set on fire.[3] As it slowly sank to the bottom, the original fleet of five ships turned into a flotilla of two.

For the journey ahead, they also needed to replenish their stocks of non-perishable food. Much like the region's residents, the crew now subsisted on a rice-based diet, but the grain sourced so far was quickly dwindling. After sailing southwest for some time, they found a suitable bay for anchorage on the western shore of Mindanao, the second largest island of the Philippines. Their hosts were friendly and willing to trade, so Carvalho bartered the longboat salvaged from the *Concepción* for supplies.[4] What had happened at Cebu apparently did not linger for long on Pigafetta's mind, as the Italian gentleman happily went exploring the region while the supplies were being put together. Pigafetta came back safe and sound,[5] but the crew was dismayed

---

[2] Fleet route adapted from Tomás Mazón Serrano ("Mapas", Ruta Elcano, June 19, 2024, https://rutaelcano.com/mapas/); map adapted from Natural Earth (1:10 m large scale land data, version 5.1.1, https://www.naturalearthdata.com/downloads/).

[3] Antonio Pigafetta, *The first voyage around the world 1519–1522*, edited by Theodore Cachey Jr. (Toronto: University of Toronto Press, 2007), on p. 63.

[4] Jean Denucé, *Magellan* [...], on p. 327.

[5] Antonio Pigafetta, *The first voyage around the world* [...], on p. 64, 65.

to find out the delivered supplies to be all perishable with no rice—the translation skills of Enrique were already sorely missed. Not all was a loss though, as the natives confirmed the itinerary to Borneo.[6]

The two ships continued travelling towards Borneo until they neared the Cagayan Sulu islands. The locals there were neither friendly nor willing to trade but, before sending away the Europeans at the song of arrows and blowgun darts,[7] they seemingly informed them about Palawan, an island to the north with supplies aplenty.[8] The intelligence was sound, and the crew succeeded in acquiring large quantities of rice and other supplies in a couple of locations along the south coast of Palawan. In the second location, probably a secluded bay today called Puerto Princesa, they overpowered a trading junk and kidnapped three of its occupants, forcing them to guide the ships to Borneo.[9]

With the aid of the kidnappees the ships finally arrived at Brunei, Borneo's major trading port. The number of houses and the amount of water traffic left no doubt that this was indeed an important commercial hub, much larger than Cebu. And, just like at Cebu, the sultan had been warned of the dealings of the Portuguese at India and Malacca, and was eager to know where the ships had come from and what they sought. Carvalho quickly put together a delegation with gifts and a message of peace from the Spanish king. Being Portuguese, Carvalho sensibly excused himself from the delegation, sending instead Espinosa, Pigafetta, and five others.[10]

Unlike the relatively down-to-earth and approachable Humabon, the Sultan of Brunei was not one to dispense with opulence and ceremony. The delegation was carried on elephants, treated to extravagant meals, and made to sleep on fine cloth. When the time came to meet the sultan, they were told they should not address the monarch directly. Instead, they needed to convey their messages to the chief of protocol, who would pass it on to a higher-ranking officer, then to an official in the anteroom, who would then use a speaking tube to deliver the message to one of the sultan's family members inside the royal room. They were also expected to show their respect to the sultan by clasping their hands above the head, raising each foot in turn, and

---

[6] Jean Denucé, *Magellan* [...], on p. 327.

[7] The Genoese Pilot in in H.E.J. Stanley, *The first voyage round the world, by Magellan. Translated from the accounts of Pigafetta, and other contemporary writers* (London: Hakluyt Society, 1874), on p. 15.

[8] Visconde de Lagôa, *Fernão de Magalhães: A sua vida e a sua viagem*, Volume II (Lisboa, Seara Nova, 1938), on p. 231.

[9] Tim Joyner, *Magellan* (Maine: International Marine, 1992), on p. 202, 203.

[10] Antonio Pigafetta, *The first voyage around the world* [...], on p. 68; Jean Denucé, *Magellan* [...], on p. 330, 331.

then blowing him kisses. The delegation found the whole setup exceptionally odd, but most probably the sultan would have reacted similarly to European court protocols.[11]

The delegation was apparently able to convince the monarch of the kindness of their motives, as the expedition was authorised to stay and conduct trade. All seemed well for about three weeks, until three armed junks anchored near the two Spanish ships, blocking their exit from the port. Then, three mariners who had been sent to shore—to purchase beeswax to caulk the ship's many leaks—failed to return. Among them was the young Joãozito, Carvalho's child from a Brazilian woman.[12] The final indication something might be amiss came when a fleet of 200 pirogues approached the Spanish ships. Fearing an attempt to seize the vessels, Carvalho ordered his gunners to attack the junks blocking the harbour entrance, killing several Malay crewmen and capturing others. Among the captured was a prince serving as a naval commander to the sultan, and three young women. The officers' plan was to trade them for their men stranded in Brunei, but it seems Carvalho had a plan of his own. According to Pigafetta, he slept with the three young virgins and made a private deal with the prince—Carvalho would let him escape, and in return the prince would negotiate the release of the hostages and secretly hand over a ransom of gold to Carvalho.[13]

It is unclear if the sultan was indeed planning to take hold of the Spanish ships, or if Carvalho was merely spooked by the sequence of events. In any case, the sultan did not react kindly to the attack and refused to hand over the stranded crewmen. After waiting for several days for a change of heart, Carvalho ordered the ships to raise anchors, leaving behind his son and four others—including two Greek mariners who, fed up with the fleet's wanderings, had deserted before the mishaps.[14]

Unable to complete the required ship repairs at Brunei, the flotilla kept close to the coast of Borneo, seeking a suitable harbour. After running aground twice but on both occasions being able to free the ships,[15] they finally managed to safely anchor further north, perhaps in the island of

---

[11] Then again, the description of these convoluted procedures is penned by Pigafetta, who was no stranger to exaggerations (Antonio Pigafetta, *The first voyage around the world* […], on p. 68–70).

[12] See Out of Portuguese Territories.

[13] Tim Joyner, *Magellan* […], on p. 206, 207.

[14] Jean Denucé, *Magellan* […], on p. 333.

[15] Antonio Pigafetta, *The first voyage around the world* […], on p. 76; The Genoese Pilot in in H.E.J. Stanley, *The first voyage round the world* […], on p. 20.

Balabac. They remained there for more than 40 days,[16] patching up as best they could the ships, both already well past their prime. A week before they left, Carvalho was relieved from command. The motion, put forward by the two remaining masters—Juan Sebastián Elcano and Juan Bautista Polcevera—was overwhelmingly supported by the crew, thoroughly fed up with Carvalho's poor hostage negotiation skills and hazardous piloting. Espinosa became the *Trinidad's* next captain, and Elcano took command of the *Victoria*.[17] Despite the new leadership and the patched ships, the expedition's key problem persisted: they still did not know the precise location of the Moluccas. After leaving Balabac, they settled on a southeasterly course, conceivably acting on intelligence gathered in Brunei.

Along the way, they looted a trading junk headed to Palawan. The junk was carrying a rajah from Palawan and his brother, who the Spaniards successfully ransomed for rice and fresh provisions.[18] This was not the first nor it would be the last act of pirating committed by the expedition. By this point, the Spaniards had learned the limitations of their weapons and armour on land and against masses of natives, but the military capabilities of their ships remained unrivalled. One can wonder if Magellan would have allowed such actions, but he probably would have. After all, the Portuguese did the same thing in the Indian Ocean, and later Francis Drake and other European corsairs would perpetrate similar state-sponsored pillages. Pirating had not been introduced by the Europeans though, as it was already an established way of life in these seas.[19]

Winds steered the flotilla southeast towards Jolo and then northeast past Basilan. Inquiring about the Moluccas, they were told about a large city in Mindanao, where perhaps further intelligence about the source of the cloves could be gathered.[20] It was a flimsy lead but the only one they had. Ironically, it was yet another act of piracy that finally gave them what they needed. On their way to Mindanao, they captured a large vessel with several chiefs aboard, one of them claiming to have been to the house of Francisco Serrão in the Moluccas.[21] Guided by the kidnapped pilot, they plotted a southeastern route. On November 6th, after capturing another pilot along the way for good measure, the flotilla finally sighted Gamalama, Ternate's prominent

---

[16] Antonio Pigafetta, *The first voyage around the world* [...], on p. 76.

[17] Jean Denucé, *Magellan* [...], on p. 334.

[18] Antonio Pigafetta, *The first voyage around the world* [...], on p. 76, 77.

[19] Tim Joyner, *Magellan* [...], on p. 210.

[20] Antonio Pigafetta, *The first voyage around the world* [...], on p. 78.

[21] The Genoese Pilot in in H.E.J. Stanley, *The first voyage round the world* [...], on p. 21.

volcano. Just as Serrão had told Magellan, the Moluccas were almost right on top of the equator.[22]

Clove trees grew on five "spice" islands: Ternate, Tidore, Motir, Makian, and Bacan, today part of the Malaku, or Moluccas Archipelago in Indonesia. Magellan may have not shared with his officers that his friend Serrão lived at Ternate, as the flotilla continued past it and anchored at nearby Tidore instead. Almanzor, the local sultan, did not consider himself bound by any previous dealings with the Portuguese. He approached the ships in a prau and warmly welcomed the Spaniards to his lands, telling them he had dreamt with the arrival of strange faraway ships. Serrão was apparently one of the reasons for Almanzor's estrangement with the Portuguese. While serving as the military advisory for Boleyse—the Sultan of Ternate—Serrão had coerced Almanzor to give his daughter in marriage to Boleyse to end a long-time dispute between the two neighbouring islands. Almazon resented the interference of Serrão and poisoned him when he came to Tidore to buy cloves.[23] This had happened seven weeks before Magellan died at Matcan: separated by less than two months and 1,100 km, the two friends had both perished half-way across the planet while meddling on somebody else's affairs.

The string of assassinations did not end there. Shortly after Serrão's death, Boleyse was poisoned by his own daughter, wife of the Sultan of Bacan.[24] Despite these local disputes, most Moluccan rulers seemed to share an animosity towards the Portuguese. This they learned from Pedro Afonso de Lorosa, a Portuguese trader living in the Moluccas for 10 years and equally disillusioned with his country. Armed with this intelligence, over the next month the Spaniards signed treaties of alliance with Tidore, Ternate, Makian, and Halmahera (the largest island of the Moluccas Archipelago, to the east of the spice islands).[25]

Despite these treaties they opted for purchasing cloves exclusively from their host Almanzor, who had committed to procuring enough to fill both ships to the brim. At the end of November, the first batch of cloves was loaded onto the ships. It was his custom, the sultan said, to properly celebrate the occasion—would the officers be so kind to join him ashore for

---

[22] Jean Denucé, *Magellan* [...], on p. 338, 339; Francisco Albo in Cristóbal Bernal, *Crónicas de la Primera Vuelta al Mundo, según sus Protagonistas*, 2016, on p. 402, 403.

[23] Antonio Pigafetta, *The first voyage around the world* [...], on p. 83–87.

[24] Idem, on p. 87, 88.

[25] José Toribio Medina, *El descubrimiento del Océano Pacífico: Vasco Núñez de Balboa, Fernando de Magallanes y sus compañeros* (Santiago de Chile: Imprenta Universitaria, 1920), on p. CCCVII, CCCVIII.

a feast? With fresh memories of the Cebu trap[26]—and having learned that three Portuguese previously employed by Serrão had been murdered at the same spring where they had been filling their water casks—they declined. Almanzor came over to the ships and, almost in tears, swore that he had no intention of breaking their alliance. He was apparently convincing, as the Spaniards agreed to stay for a couple more weeks and leave five men behind in a trading post.[27] Amidst quarrels with neighbouring islands and after assassinating at least four Portuguese, Almazar did seem to have good reasons not to gain additional enemies.

Finally, on December 18th, fully loaded with cloves and provisions, the *Trinidad* and the *Victoria* were ready to raise anchors. Their new sails carried freshly painted crosses of Santiago, with a hopeful inscription below: *"This is the sign of our good fortune"*.[28]

14.1 What Happened to the Moluccas?

The short answer

Portugal promptly complained about Spain's presence in the Moluccas, and a multi-year diplomatic battle ensued. In 1529, with neither country able to prove the location of the Moluccas in relation to the Tordesillas anti-meridian, the Spaniards sold their claims to the Portuguese for 131 million maravedies, enough to finance Magellan's expedition 13 times over. In the ensuing decades, Portugal likely recouped the treaty's payout and then some, but their stronghold in the region began to falter in the 1570 s, first from clashes with local powers and then from the arrival of the Dutch. By the mid-sixteenth century, the Portuguese had been ousted from the region and the Dutch controlled the spice islands. During more than a century, they avoided overproduction and smuggling by keeping clove trees under a watchful eye. The Dutch hegemony ended in 1749, when a Frenchman managed to pilfer clove seedlings and acclimate the plant to Mauritius. Today, clove grows in several tropical countries, but the Moluccas and other parts of Indonesia continue to be the largest producers. However, the once eye-watering expensive spice is now mostly used to make kreteks, a clove flavoured tobacco.

---

[26] See The Ambush.

[27] Antonio Pigafetta, *The first voyage around the world* [...], on p. 94, 95.

[28] Idem, on p. 98.

News of the arrival of Magellan's fleet to the Moluccas travelled swiftly to Portugal but not fast enough to reach the ears of Manuel I, who had perished from high fevers during an outbreak of Black Plague.[29] It was John III, his successor, who complained to the Spanish king that the Treaty of Tordesillas had been breached.[30]

After a couple of years of unfruitful back-and-forth letters and audiences with ambassadors, the two countries agreed to meet at a bridge over Caia, a river bordering the Spanish town of Badajoz and the Portuguese one of Elvas. The several dozens of attendees were divided into two groups: astronomers and mariners would discuss the *ownership* of the Moluccas, a scientific matter depending on the position of the islands relative to the demarcation line; while lawyers would debate their *possession*, a legal issue hinging on who had arrived there first.[31]

Each party carefully instructed their attendees on the strategy to follow. On ownership, Spain would refer to Ptolemaic geography and Portugal's own charts, both of which suggested the Moluccas fell on their hemisphere. The Spaniards' argument for possession was more devious but no less effectual: rather than admitting they knew Portugal had arrived at the Moluccas ten years ago, they would claim Magellan's fleet had got there first, placing the onus of proving otherwise on the other party.[32]

Portugal, on the other hand, wanted to quickly quell any discussion about ownership by dismissing charts as inadmissible evidence, as several of the country's mathematicians and astronomers had already expressed concern about distortions. In alternative, they declared the astronomical measurement of longitude to be the only trustworthy way of locating the Moluccas. As this would require time, they hoped to shift the discussion to the issue of possession, where they could bring in evidence of their expansion in Southeast Asia.[33]

The discussions dragged on for a couple of months before ending in a deadlock, with neither party capable of breaking the opposing arguments. Two key pieces of information, perhaps capable of shifting the argument

---

[29] Edward McMurdo, *The history of Portugal, from the Commencement of the Monarchy to the Reign of Alfonso III* (London: Sampson Low, Marston, Searle, & Rivington, 1889), on p. 115.

[30] "Minuta da carta de D. João III a Luís da Silveira a respeito do que diria ao imperador Carlos V sobre a nau que viera de Maluc" (Lisbon: Arquivo Nacional da Torre do Tombo, PT/TT/GAV/15/1/59, 1522).

[31] An English translation of the negotiation's proceedings and related correspondence is available at Emma Blair and James Robertson, *The Philippine Islands*, Volume 1 (Cleveland: A.H. Clark Company, 1903), on p. 145–221.

[32] Idem.

[33] Idem.

to Portugal's court, never made it to that bridge: San Martín's longitude measurement at Suluan[34]; and Pigafetta's diary, which contained ample mentions to the Portuguese presence in the islands.[35]

Neither country was interested in a military escalation, so the deadlock lingered. The impasse favoured Portugal, who continued to benefit from the lucrative trading of cloves. Over the following years, Spain sent several follow-up expeditions to the Moluccas, but only a trickle of ships succeeded in retracing the dangerous route Magellan had trail-blazed. Most of those that did eventually fell into the heads of the Portuguese,[36] who leveraged their Malaccan hub to maintain a strong presence in the Moluccas.

In 1525, João III married Charles V's sister, and in 1528 the Spanish king married the Portuguese monarch's sister, deepening the bond between the two kingdoms and placing additional pressure on the need to find a diplomatic solution for the spice standoff. In 1529, they finally reached an agreement, set into paper by the Treaty of Zaragoza. Neither party was willing to concede their previous arguments, so the solution was to throw money at the problem. And a lot of it, as Portugal had to pay 350,000 ducats of gold[37] (131 million maravedies) in exchange for Spain's relinquishment of any claims to the Moluccas.

It was a lavish amount of gold. Still, the Spaniards claimed it was less than half of the true value of the Moluccas, a show of goodwill towards their neighbours.[38] It was Spain that arguably made a sugary deal though, as only about 1/8 of the clove produced in the Moluccas was sent directly to Europe,[39] with the bulk of the remaining volume traded in Malacca, a port Portugal controlled. Even if they could increase the amount delivered directly to Europe, they still lacked a viable return route, as their first successful Pacific eastward crossing only took place in 1565. With the treaty, Charles V was pocketing more than 13 times what Magellan's expedition had cost,[40] quite a handsome profit.

---

[34] See Were San Martín's Longitude Measurements Accurate?

[35] See The Moluccas.

[36] See What Happened to the Philippines?

[37] Emma Blair and James Robertson, *The Philippine Islands* […], on p. 224.

[38] Idem, on p. 238.

[39] Luís Thomaz, "Maluco e Malaca", *Actas do II Colóquio Luso-Espanhol de História Ultramarina*, 1975, p. 27–48, on p. 48.

[40] Based on a total expedition cost of ~10.1 m maravedies (Lourdes Díaz-Trechuelo, "La organización del viaje Magallânico", *Actas do II Colóquio Luso-Espanhol de História Ultramarina, 1975*, p. 265–314, on p. 287).

For Portugal, the bitter pill was perhaps easier to swallow as the deal placed the demarcation line 17° east of the Moluccas,[41] giving the small country a hemisphere of influence of 192° and reducing the slice of its neighbour to the remaining 168°. Apart from the satisfaction of keeping more than half of the globe, the most immediate opportunity offered by the extra 17° was the inclusion of the Banda Islands, the source of nutmeg and mace.[42]

The treaty was a pawn rather than a sale, as Spain could revert to the previous *status-quo* by returning the gold, and Portugal could also get it back by proving the Moluccas was less than 180° east of the Tordesillas meridian.[43] Neither signatory ever triggered these clauses and were at times quite lenient with transgressions. For instance, Portugal did not protest too vigorously when Spain occupied the Philippines, and Spain allowed Portugal to expand Brazil westward well beyond the Tordesillas meridian.

The agreement ended the local scuffles between Spain and Portugal, allowing the latter to strengthen its presence in the Moluccas, with Ternate serving as its main hub. The relations with the natives went through ups and downs but took a definitive turn to the worst when a Portuguese killed the sultan of Ternate in 1570. The locals rebelled and, with the aid of some of the surrounding sultanates, expelled the occupiers in 1575, who retreated to Ambon, in the Banda Islands. The trade of clove faltered but did not break, as the Portuguese had acclimated clove trees to Ambon. In 1578, the sultan of Ternate tried to reduce his Tidore counterpart to a vassal, prompting the latter to accept the Portuguese back in exchange for military protection.[44]

Back in Europe, changes were also brewing. Portugal and Spain's monarchs had kept close blood relations, and the untimely death of the young Portuguese King Sebastião I in 1578 meant the next in line for the throne was Philip II, Charles V's son and by then the Spanish king. This union of the Crowns led Portugal to also sever commercial ties with the Netherlands when they fought Spain for independence,[45] prompting them to go looking for the source of cloves. By 1599, they had reached the Moluccas. Over the next decades, the combined forces of Spain and Portugal fought to maintain

---

[41] Emma Blair and James Robertson, *The Philippine Islands* […], on p. 226.

[42] Juan Carlos Rey, "Clavo y nuez moscada", *El viaje más largo: La primera vuelta ao mundo* (Madrid: Sociedad Mercantil Estatal de Acción Cultural, 2019), on p. 234.

[43] Emma Blair and James Robertson, *The Philippine Islands* […], on p. 225, 227, 228.

[44] M. C. Ricklefs, *A History of Modern Indonesia since c. 1300* (Houndmills: Macmillan, 1991), on p. 25.

[45] See What Happened to the Philippines?

control, but the presence of the Dutch—well organised and well equipped—steadily increased. By the middle of the seventeenth century, the Iberians were gone from the Moluccas.[46]

Despite the resistance of the local kingdoms and the occasional brawl with the British here and there,[47] the Dutch colonies in the East continued to expand, eventually occupying most of modern-day Indonesia. During that time, the spice left the original spice islands. The Dutch eradicated clove trees from the Northern Moluccas to limit production to the Banda Islands, where the Portuguese had already acclimated cloves and which were easier to defend.[48]

The Netherlands intended to avoid overproduction and smuggling, which they managed to do for more than 150 years. Cloves were hard to smuggle because they are a dried flower, rather than a fruit or seed which can be planted, and its seedlings would usually not survive transport. In 1749, the French Crown tasked the multitalented Pierre Poivre with breaking the Dutch monopoly of clove and nutmeg. The mission, punishable by death, was not for the faint of heart but Poivre set to it with purpose. After more than two decades and several failed attempts, Poivre was finally able to acclimate cloves and nutmeg to the French colony of Mauritius, using nutmeg trees and clove seedlings taken—quite literally—from under Dutch noses. The *Vigilant*, the French ship carrying the smuggled goods, had been hailed by a patrol boat while still in Dutch waters. Jean Provost, the navy officer in charge, claimed they had been blown off course by a storm. The officers bought the excuse, declining to search the vessel and even escorting it out of the surrounding reefs.[49]

Despite the acclimation of clove trees to other countries, today Indonesia still produces more than 70% of the world's cloves. Yet, things have changed. Clove farmers are no longer exclusive to the Moluccas, and many grow it instead in Sulawesi and Java. And, rather than seasoning dishes, 90% of Indonesia's cloves are used to make *kreteks*, a spice flavoured tobacco.[50] *Kreteks* are a striking example of globalisation, as its main ingredients come literally from opposite sides of the globe.[51]

---

46 Luís Thomaz, "Maluco e Malaca" […], on p. 39.

47 M. C. Ricklefs, *A History of Modern Indonesia since c. 1300* […]„ on p. 28, 29.

48 Luís Thomaz, "Maluco e Malaca" […], on p. 39.

49 Charles Corn, *The scents of Eden: A narrative of the spice trade* (New York: Kodansha International, 1998), on p. 219–225.

50 Patricio Marquez et al., *The economics of clove farming in Indonesia*, (Washington D.C.: World Bank Group, 2017), on p. 4.

51 The antipode of the Moluccas is in South America, from which the tobacco plant originally came from (see How Did 16th Century Exploration Change Food Habits Around the Globe?).

15

The Eastern Dead-End

From the Moluccas (April 6th, 1522) and Back to the Moluccas (Late October, 1522)

The new "*this is the sign of our good fortune*" insignias[1] quickly proved optimistic for the *Trinidad*. Once the flagship of a proud fleet of five ships, now it could not even make it out of the harbour. The strain of raising the anchors was enough to breach the overloaded and weakened hull, and water quickly started pouring in. The *Victoria* came back to see what happened, only to find their comrades frantically operating the bilge pumps while trying to locate the breach. Almanzor sent specialised divers with long hairs to pinpoint the location of the hull breach, but they too failed.[2]

After a while it became clear the *Trinidad* would require extensive repairs to make it seaworthy again. At least the incident had happened on a friendly port, and Almanzor reportedly put at their disposal more than 200 workers. Time was of the essence though, as the window of favourable easterly winds was quickly closing.[3] This left the Europeans with three options, none of them particularly compelling. Option one was to conduct the repairs on the *Trinidad* and then wait for the next season to sail west, greatly increasing the risk of an encounter with the Portuguese. Option two was to sail east once the repairs were done and try to reach a Spanish settlement in the west coast of the Americas.[4] This course of action would minimise the chances

---

[1] See The Moluccas.

[2] Antonio Pigafetta, *The first voyage around the world 1519–1522*, edited by Theodore Cachey Jr. (Toronto: University of Toronto Press, 2007), on p. 100, 101.

[3] Tim Joyner, *Magellan* (Maine: International Marine, 1992), on p. 218.

[4] Antonio Pigafetta, *The first voyage around the world 1519–1522* [...], on p. 101.

R. Gaspar, *The Revolution of Magellan*, https://doi.org/10.1007/978-3-032-10797-8_15

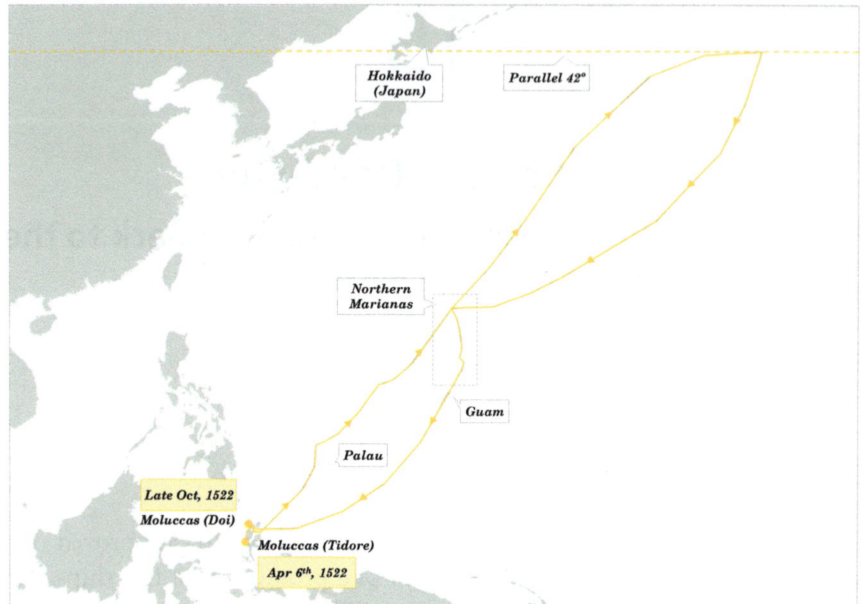

Fig. 15.1 The *Trinidad* route[5]

of finding Portuguese ships, at the expense of crossing again the Pacific. The third and final option was to hedge their chances: the *Victoria* would set sail immediately to the west, and the *Trinidad* would sail to the east once it could. An agreement was eventually reached to follow this third option. It was hard to judge who had drawn the shortest straw. The crew of the *Victoria* would need to round the Cape of Good Hope to get back to Spain, a clear breach of the Treaty of Tordesillas. The *Trinidad*, in exchange for causing less of a diplomatic snafu, needed to venture again into unchartered waters (Fig. 15.1).

On December 21st, the *Victoria* set sail with a lightened cargo of cloves—to avoid what had happened to the *Trinidad*—and carrying the many letters of those staying behind wrote to their loved ones. Pigafetta was part of the party, although it is unclear when the Italian rapporteur made the switch from

---

[5] Fleet route adapted from Tomás Mazón Serrano ("Mapas", Ruta Elcano, June 19, 2024, https://rut aelcano.com/mapas/); map adapted from Natural Earth (1:10 m large scale land data, version 5.1.1, https://www.naturalearthdata.com/downloads/).

the *Trinidad* to the *Victoria*, or if he had a say in the matter.[6] In any case, the move may have saved his life.

It took almost four months to make the *Trinidad* watertight again. When it finally set sail, on April 6th, the *Trinidad* carried a crew of 54. These were nearly all the Europeans still at Tidore, except for five left behind to guard a trading post with excess goods. Lorosa, the talkative Portuguese trader who had helped the Spaniards,[7] decided to try his luck aboard a Spanish ship and joined the *Trinidad*'s crew. Espinosa remained its captain, piloting duties were assumed by two Genoese (Giovanni Battista di Polcevera, the *Trinidad*'s original master, assisted by Leone Pancaldo, a sailor), and the sailor Diego Martín was promoted to master.[8] Their strategy was to sail north until they found westerly trade winds to carry them across the Pacific. It was a sound plan in theory, as they knew such winds existed in the North Atlantic and correctly assumed the same dynamics for the North Pacific. However, they were not aware of the monsoons in the Philippine Sea which, from January to June, blow from the northeast and disrupt trade wind patterns.[9]

They encountered the monsoon winds shortly after sailing away from the Moluccas, having no other option than to resort to tacking against the wind. They slowly and labouredly progressed northeast, sighting islands from Palau and Micronesia.[10] On June 11th, the *Trinidad* anchored on one of the Northern Mariana Islands. As in Guam, the ship was quickly encircled by curious indigenous. This time however the Europeans did not chase them away violently and established instead a friendly rapport, so much so that one of the natives joined them as a local pilot.[11] They were now making better progress towards the north, aided by the approaching summer—which brought winds blowing from the east, rather than the northeast—and the Kuroshio Current.[12]

Despite the good progress, there was still no hint of the westerlies. Days and nights became increasingly harder on the crew as the temperature dropped, the supplies dwindled, and more men became incapacitated by disease. By this time the *Trinidad*'s crew was undoubtedly recalling the plights

---

[6] Tim Joyner, *Magellan* [...], on p. 218.

[7] See The Moluccas.

[8] José Toribio Medina, *El descubrimiento del Océano Pacífico: Vasco Núñez de Balboa, Fernando de Magallanes y sus compañeros* (Santiago de Chile: Imprenta Universitaria, 1920), on p. CCCXXIX, CCCXXX.

[9] Tim Joyner, *Magellan* [...], on p. 221.

[10] José Toribio Medina, *El descubrimiento del Océano Pacífico* [...], on p. CCCXXXII.

[11] Jean Denucé, *Magellan: la question des Moluques et la première circumnavigation du globe* (Brussels: Hayez, 1911), on p. 369.

[12] Tim Joyner, *Magellan* [...], on p. 222.

they had endured in Patagonia and in the Pacific. To make matters worse, the ship was hit by a storm so severe it tore off both the fore and aft castles, broke the main mast, and ripped the sails to shreds. At an estimated latitude of 42°—the same parallel as Hokkaido, Japan's northernmost main island—the *Trinidad's* officers decided to retreat to the Mariana Islands.[13]

While they had gone sufficiently north to catch the westerly trade winds, they had missed the time window to do so. The later Spanish galleons running the yearly route from Manila to Acapulco were often not allowed to leave the Philippines after June, as the conditions deteriorated significantly afterwards. Even then, many returned to Manila after facing strong winds and waves off the coast of Taiwan and Japan. Those who made it close to the 40° parallel—the latitude at which they could hop onto the westerly trade wind that would take them across the Pacific—still had a long journey ahead, packed with temperamental weather, disease, and pirates. It would still take Spain nearly 50 years to successfully cross the Pacific from the west to the east, with the *Trinidad* merely being the first of several failed attempts.[14] Giovanni Careri, a 17th-century Italian gentleman who, not unlike Pigafetta, travelled for adventure rather than profit, wrote:

> The voyage from the Philippine Islands to America may be called the longest and most dreadful of any in the world, as well because of the vast ocean to be crossed being almost one half of the terraqueous globe, with the wind always ahead, as for the terrible tempests that happen there, one upon the back of another, and for the desperate diseases that seize people in seven or eight months, lying at sea sometimes near the line, sometimes cold, sometimes temperate, and sometimes hot, which is enough to destroy a man of steel, much more flesh and blood, which at sea had but indifferent food.[15]

On their way back, the *Trinidad* briefly stopped at a tiny island in the Northern Marianas, perhaps Maug.[16] Their local pilot went ashore and came back with some much-needed fresh provisions, but the island's meagre resources could not sustain the entire crew. After filling some casks with the

---

[13] Jean Denucé, *Magellan* [...], on p. 369, citing António Galvão.

[14] See What Happened to the Philippines?

[15] Giovanni Careri in Arturo Giraldez, *The Age of Trade: The Manila Galleons and the Dawn of the Global Economy* (Lanham: Rowman & Littlefield, 2015), on p. 119.

[16] The Genoese pilot (likely Leon Pancaldo) called it "Mao" (H.E.J. Stanley, *The first voyage round the world, by Magellan. Translated from the accounts of Pigafetta, and other contemporary writers*, London, Hakluyt Society, 1874, on p. 28).

water they could find, they decided to sail straight back to the Moluccas rather than spend more time searching for supplies. Four sailors apparently did not think this was the wisest decision and deserted. Espinosa convinced one of them to come back, but the remaining three would not even come near the battered ship and sickly crew.[17] In 1526, the Loaísa expedition would find one of them at Saipan—the larger of the Northern Mariana Islands—who reported the other two had been killed by natives.[18] Their chances of survival would have been only marginally better if they had stayed aboard though. When the *Trinidad* finally limped back to the Moluccas, after a back-and-forth journey of more than seven months, 33 men—over 60% of the crew—had died.[19] Most of them perished from scurvy, but a mysterious disease also seemed to be roaming the decks. Puzzled by its symptoms, the crew performed a makeshift autopsy on one of the deceased, finding a parasite in his stomach.[20]

Rather than sailing back to Tidore, the *Trinidad* stopped at Halmahera, the nearby large island ruled by a friendly sultan with whom they had established a peace treaty. Down to only seven men strong enough to man the ship and a single anchor to hold it steady, the situation was dire. Locals also informed Espinosa that while they were away a fleet of seven Portuguese ships had arrived at Ternate, in pursuit of the Spaniards. Espinosa apparently decided it was useless to fight or flight the Portuguese, and asked them instead for help. Bartolome Sanchez, the *Trinidad's* clerk, was sent to Tidore with a desperate plea for provisions and an extra anchor. Any hope for some Christian brotherhood ended when the Portuguese arrived, promptly arresting the crew and seizing the ship. Everything aboard, from the cargo of cloves to the expedition's logbooks, was confiscated. Head held high, Espinosa asked for a notarised list, but the only thing he received in return was a threat he would get the list while hanging from a yard. Lorosa was executed for treason, and the *Trinidad* would later ground and shatter to pieces during a storm, finally putting the once proud flagship out of its misery.[21]

---

[17] José Toribio Medina, *El descubrimiento del Océano Pacífico* […], on p. CCCXXXIV.

[18] Idem, on p. CCCCLIII, CCCCLIV.

[19] The tally, made by Medina based on available documentation, is for the European crew only (José Toribio Medina, *El descubrimiento del Océano Pacífico* […], on p. CCCXXXV). The fate of the 13 Moluccans who also sailed on the *Trinidad* was poorly documented and unclear.

[20] José Toribio Medina, *El descubrimiento del Océano Pacífico* […], on p. CCCXXXIII, citing Herrera.

[21] José Toribio Medina, *El descubrimiento del Océano Pacífico* […], on p. CCCXXXVI–CCCXXXVIII; Jean Denucé, *Magellan* […], on p. 371, 372.

15.1   Who Were the First to Cross the Pacific?

The short answer

Recent genetic analyses suggest Polynesians and Native Americans met in the Marquesas Islands around the 12th century. So, some 400 years before Magellan crossed the Pacific, either Native Americans travelled to the Marquesas and settled there, or Polynesians sailed to and from the Americas, bringing back a few Native Americans with them. Nobody is quite sure which of these two voyages happened: Polynesians were far more experienced in navigation, but a return trip to the Americas would have been an immense challenge; Native Americans were not known for long-distance sailing, but a lost raft could conceivably have drifted to the Marquesas. Who then crossed the Pacific first? Magellan and his crew were most likely the first to ever cross the entire Pacific Ocean in the western direction, because both theories only take the Polynesians and Native Americans to the Marquesas, less than half of the length of this ocean. The first ever Pacific crossing in the eastward direction is unclear. If Polynesians did indeed reach the Americas, then they were the first to do so, although in a piecemeal fashion. The first non-stop crossing was only accomplished several centuries later, in 1565, with the Spanish Manila to Acapulco route.

Assuming Magellan and his crew were the first to cross the Pacific seems reasonable. The fleet barely made it despite using state-of-the-art 16th-century ships and navigation techniques,[22] so the likelihood somebody else beat them to the punch is small. Yet, a nagging detail does not quite fit that theory: Polynesians were already eating sweet potatoes long before Columbus first brought them from the Americas.

The most straightforward explanation is that sweet potato seeds drifted across the Pacific to the Polynesian islands. These seeds are remarkably stout and will germinate even after being submerged in saltwater for 120 days, roughly the time required for them to travel from the Americas to the Marquesas Islands pushed by the South Equatorial current. Long-distance dispersal is not uncommon and has been confirmed for close relatives of sweet potatoes. Also, DNA analysis of an early specimen collected in Polynesia suggests it diverged from American sweet potatoes more than 100,000 years ago, well before any humans settled the Americas and Polynesia.[23]

Case closed, then? Not quite. The pre-human settlement dating of Polynesian sweet potatoes is based on a single specimen, collected more than

---

[22] See How Advanced Were Magellan's Ships? and What Navigation Techniques Were Available to Magellan?

[23] Pablo Muñoz-Rodríguez et al., "Reconciling Conflicting Phylogenies in the Origin of Sweet Potato and Dispersal to Polynesia", *Current Biology*, 28(8), 2018, p. 1246–1256.e12, on p. 1253, doi:10.1016/j.cub.2018.03.020.

250 years ago and stored in the British Museum. Dating ancient DNA is notoriously fiddly and prone to errors (as DNA starts to degrade as soon as the organism dies) so additional specimens would be needed to confirm the results. The hypothesis of natural dispersal also does not explain the similarities between the Quechua word for sweet potato—"*cumal*"—and the Polynesian one—"*kumara*".[24]

Sparked by this possible linguistic connection, discussion of early contact between Native Americans and Polynesians has been going on for decades. In the 1990s, not long after DNA profiling was first used to solve crimes,[25] it was also brought into this conundrum. Initial DNA tests focused on Rapa Nui (Easter Island), as this middle-point island seemed the most probable place for an encounter between Polynesians arriving from the west and Native Americans coming from the east. These early tests, based only on a small number of samples, were inconclusive. Native American DNA was indeed found on some of the samples, but it was not clear if the admixture had happened before or after Easter Island was annexed by Chile in the nineteenth century.[26]

In 2020 further tests were made, taking advantage of the development of advanced genome analysis techniques. Rather than focusing on Easter Island, DNA samples from more than 800 individuals belonging to 32 Polynesian and Native American groups were tested. The results were fascinating: one, contact between Polynesians and Native Americans seems to have taken place before the arrival of Europeans, around the twelfth century; two, contact happened not in Easter Island but in the Marquesas Islands; three, it was likely a single event, rather than recurrent contacts over time; four, the Native American DNA most closely matched the Zenú people, who lived in coastal present-day Colombia.[27]

These results are based on DNA from living individuals and need to be confirmed with ancient DNA testing. At this point though, it seems Polynesians and Native Americans did in fact meet each other well before the arrival of Europeans. But who travelled across the Pacific? Did Polynesians reach the

---

[24] Lisa Matisoo-Smith and Michael Knapp, "When did sweet potatoes arrive in the Pacific—Expert Reaction", Science Media Centre, 2018, https://www.sciencemediacentre.co.nz/2018/04/13/when-did-sweet-potatoes-arrive-in-the-pacific-expert-reaction/.

[25] DNA was first used in forensic science in 1986, in the United Kingdom (S. Panneerchelvam and M.N. Norazmi, "Forensic DNA Profiling and Database", *Malaysian Journal of Medical Sciences*, 10(2), 2003, p. 20–26, on p. 22).

[26] Alexander Ioannidis, Javier Blanco-Portillo et al., "Native American gene flow into Polynesia predating Easter Island settlement", *Nature*, 583, 2020, p. 572–577, on p. 577, doi:10.1038/s41586-020-2487-2.

[27] Idem.

Americas and brought back Native Americans, or did Native Americans travel to the Marquesas and established themselves there?

Polynesians were undoubtedly more familiar with the seas than any of the Native American peoples. Over a period of more than 4,200 years (Fig. 15.2), Polynesians settled islands across 155 degrees of longitude, from Madagascar to Rapa Nui (Easter Island), and across nearly 70 degrees of latitude, from Hawaii to Aotearoa (New Zealand). Polynesians did not use charts nor nautical instruments and their canoes used only Stone Age techniques, leaving Europeans explorers baffled on how they had succeeded in settling such a large territory. Theories to explain such a feat varied widely. Late 16th-century explorers expected to find a chain of closely spaced islands connecting the remote Pacific islands to mainland Asia. Some believed God had first created Polynesians as a single people and then scattered them across the isolated island groups. Others postulated the existence of a vast ancient continent that had once connected all the Polynesian islands, long gone due to some sort of cataclysm. These theories lost credibility over time, but many were still left believing Polynesians had found new islands entirely by chance. The likelihood of drifting into an island in the middle of the ocean is small—Magellan crossed the entire Pacific sighting only a couple of islets[28]—but the theory endured well into the 1960s, until traditional long-distance Polynesian voyages were successfully recreated.[29]

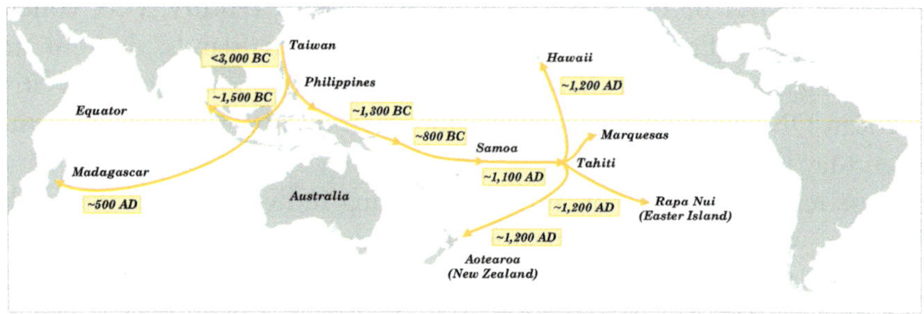

Fig. 15.2 Approximate dating of Polynesian expansion[30]

---

[28] See Why Did Magellan Take Such a Convoluted Route Across the Pacific?

[29] Ben Finney, *The History of Cartography* (Chicago: The University of Chicago Press, 1998), in Volume Two, Book Three, p. 444.

[30] Adapted from Peter Bellwood et al., "Are 'Cultures' Inherited? Multidisciplinary Perspectives on the Origins and Migrations of Austronesian-Speaking Peoples Prior to 1000 BC", *Investigating Archaeological Cultures*, edited by Benjamin Roberts and Marc Vander Linden (New York: Springer, 2011), on p. 340, https://doi.org/10.1007/978-1-4419-6970-5_16; Janet Wilmshurst et al., "High-precision radiocarbon dating shows recent andrapid initial human colonization of East Polynesia", *PNAS*, 108(5), 2010, p. 1815–1820, on p. 1816, doi:10.1073/pnas.1015876108.

The scepticism surrounding Polynesian navigation was not only caused by how different it was from European practices, but also because it was never fully documented. Rather than written, these methods were transmitted orally from the elders to the young. The arrival of Europeans brought diseases which debilitated the native population and new navigation techniques which replaced the traditional methods before these had been thoroughly recorded. Much of what is known today comes from the Caroline and Marshall Islands—the only two archipelagos where traditional navigation continued to be practised after the European arrival—and solitary fragments from other Polynesian islands.[31] The resulting picture, while fuzzy at points, provides a glimpse into how Polynesians found their way across long distances.

These navigators never seemed to use charts, even for month-long island crossings. Pictorial representations—using sticks and shells—were only used on land, either to train new navigators or to show to others where surrounding islands were located. At sea, pilots relied on information they had memorised, organised around a "mental compass rose" (Fig. 15.3).

Polynesian pilots got their bearings from where the wind was blowing (keeping in mind prevalent winds changed with seasons) and from where stars were raising and setting on the horizon. During the day and cloudy nights, the pilot would get guidance from where the sun raised, reached its

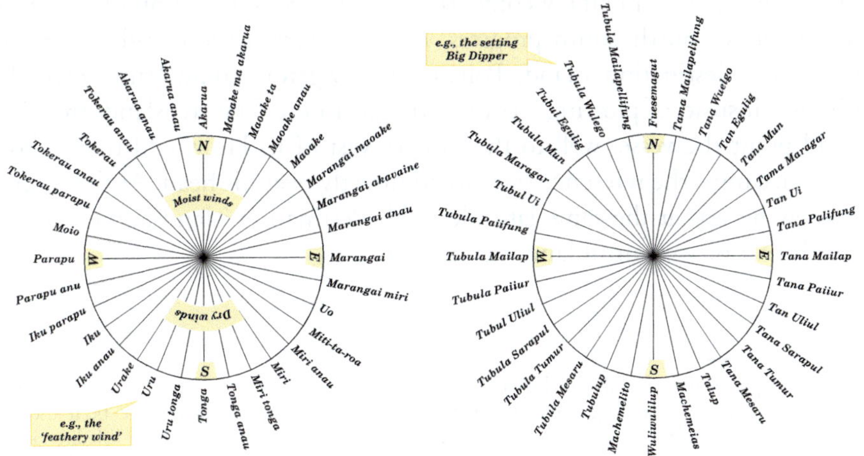

Fig. 15.3   Left: example of a wind mental compass rose from the Cook Islands; right: example of a star mental compass rose from the Caroline Islands[32]

---

31 Ben Finney, *The History of Cartography* […], on p. 454.

32 Adapted from Ben Finney, *The History of Cartography* […], on p. 459, 462.

peak, and then set (remembering that these change throughout the year), and from the direction of currents and swells.[33]

By superimposing the position of known islands onto these mental compasses, pilots were able to draw the approximate course needed to go from one island to another. On short intra-archipelago trips, progress could be easily estimated by picking a reference island and checking which mental compass direction was directly behind it (Fig. 15.4). On longer open sea voyages with no islands in sight, the pilot's reference island was not a visible landmark but rather a mental one. This concept was not unlike European dead reckoning,[34] albeit with two key differences: one, bearings were not measured with a compass but instead estimated by comparing the direction of the canoe with that of winds and stars; two, instead of calculating and plotting the vessel's progress onto a chart, pilots kept a mental note of how many *etak* (route segments crossed by two adjacent compass directions) have been sailed. To improve the accuracy of this mental process, experienced navigators cross-referenced the direction of the winds and stars with further data points such as currents, swells, bird migration patterns, and cloud formations.[35]

Much like for European dead reckoning, the errors of Polynesian dead reckoning quickly compounded during longer voyages, particularly when unfavourable winds had forced the vessel off-course. To mitigate this issue, it seems some Pacific islanders also resorted to an instrument-less version of latitude navigation, mentally converting the change of elevation of stars into an estimate of north–south progress. If a star crossed the meridian directly above their destination island, Polynesian navigators could "run down the latitude": instead of plotting a course straight to the faraway island, the pilot would point the vessel well to the east or west of it. He would then follow the course until the reference star passed directly over his head. At that point, he knew he was at the same latitude as his destination island and could sail directly east or west until land was sighted.[36]

---

[33] European portolan charts also included the names of predominant winds on the cardinal directions of compass roses, leading some to theorise that mental compasses may have been used in the Mediterranean before the introduction of the magnetic needle (Tony Campbell, "Mediterranean portolan charts: their origin in the mental maps of medieval sailors, their function and their early development", Map History, 2021, https://www.maphistory.info/PortolanOriginsTEXT.html#note80).

[34] See What Navigation Techniques Were Available to Magellan?

[35] Ben Finney, *The History of Cartography* [...], on p. 459–485.

[36] European navigators also *ran down the latitude,* although latitude was measured astronomically instead of visually estimated (see What Navigation Techniques Were Available to Magellan?).

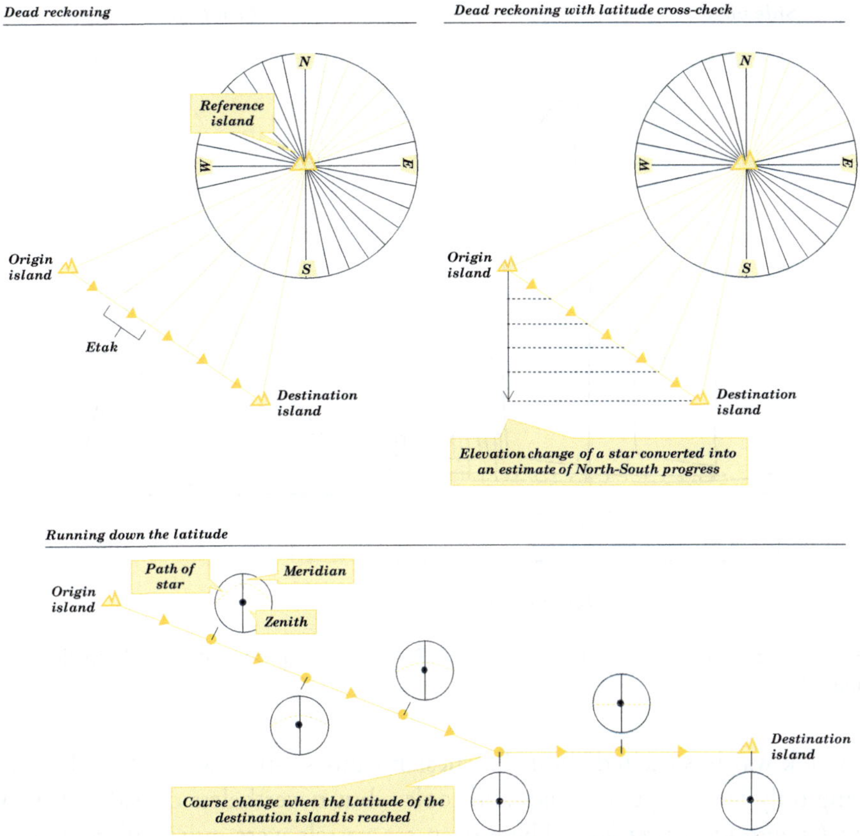

Fig. 15.4 Examples of three Polynesian navigation techniques: dead reckoning (upper left), dead reckoning with latitude cross-check (upper right), and running down the latitude (bottom)[37]

Twentieth-century voyages recreated some of these traditional Polynesian navigational methods, confirming they can indeed bring vessels to good port across long voyages. Before, in the eighteenth century, the British explorer James Cook had already expressed admiration for these techniques. While visiting Tahiti, Cook met Tupaia, an elder who advised the high chiefs in matters of geography, meteorology, and navigation. Tupaia joined Cook aboard the *HMS Endeavour* and astonished the British pilots with his navigation skills in a long journey across New Zealand, Australia, and Java.[38]

---

[37] Adapted from Ben Finney, *The History of Cartography* [...], on p. 460, 471–475.

[38] Ben Finney, *The History of Cartography* [...], on p. 448, 449.

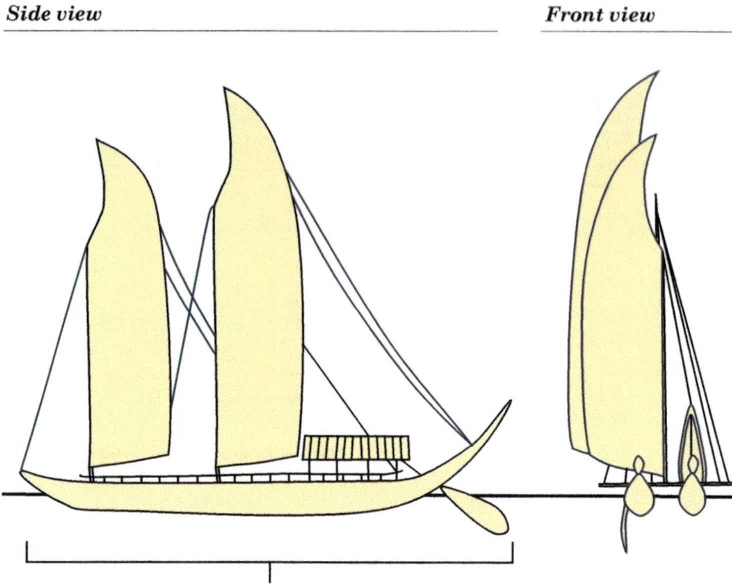

Fig. 15.5 Example of a Tahitian *pahi*, a Polynesian canoe used for long-distance sailing[41]

Cook was less excited about Polynesian canoes—they were, after all, built using only Stone Age materials and tools—but nevertheless considered them "*fit for distant navigation*". These deep-sea vessels were quite different from the small and nimble outrigger canoes that had surrounded Magellan's fleet in Guam.[39] They were double-hulled (Fig. 15.5)—providing stability in strong winds and additional cargo capacity—and large—15 to 22 m, not much less so than the *Trinidad.* The hulls were made of planks fastened to a keel and ribs using coconut fibres. The lack of bolts and nails made for a leaky boat—requiring frequent water bailing—but also meant it would bend rather than break under a heavy storm.[40]

Together, these navigation methods and deep-sea canoes enabled Polynesians to settle the eastern limits of the Pacific. The best documented long-distance course is between Tahiti and Hawaii, which are about 4,000 km apart. Tahiti may also have been the hub from where the Easter Island (~4,300 km away) and New Zealand (~3,900 km) were settled. Some believe

---

[39] See Bad News.

[40] David Lewis, *We, the Navigators: The ancient art of landfinding in the Pacific* (Canberra: Australian National University Press, 1965), on p. 272.

[41] Adapted from David Lewis, *We, the Navigators* […], on p. 256, citing Jean Neyret.

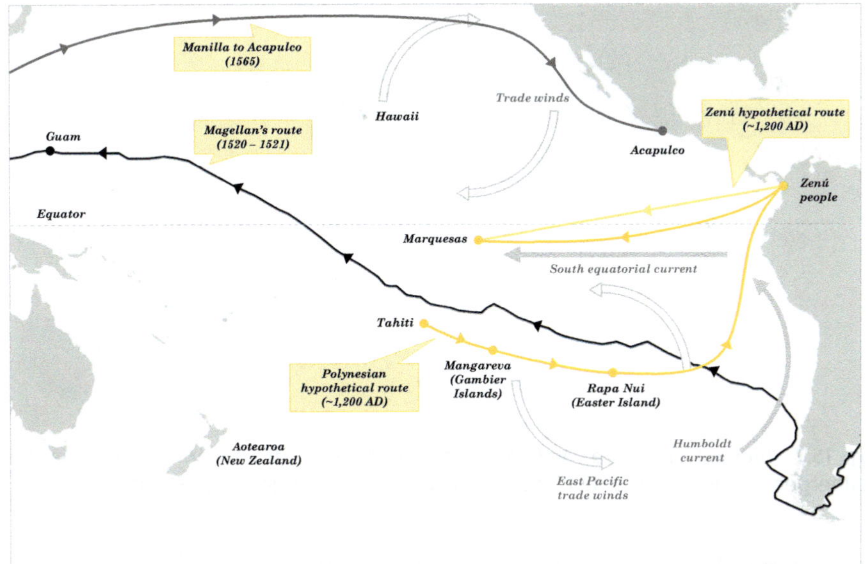

Fig. 15.6 Comparison of the hypothetical 13th-century Polynesian and Native American Pacific crossings with those achieved in the sixteenth century[43]

longer stretches of 8,400 km to be entirely feasible (six weeks at sea, doing about 200 km per day) if the winds were favourable and provisions plenty.[42] This sort of range puts a return trip to the Americas within reach of a well-prepared and knowledgeable crew (Fig. 15.6).

Still, luck would need to be on their side. A potential route would start in Tahiti and make use of the trade winds to make steadfast eastern progress, perhaps with reprovisioning stops at Rapa Iti, Mangareva (Gambier Islands) or Rapa Nui (Easter Island). Once they reached the South American coast, the Humboldt Current could take them north up to coastal Colombia and in contact with the Zenú people. After such a long journey, the Polynesian expedition would be in dire need of supplies and ship repairs. These were unfamiliar waters for them, so they could also need help from the natives to find favourable winds and currents to return west. Plus, to explain the combined Polynesian and Native American DNA, some Zenú would need to join the journey back home, either willingly or unwillingly. The most probable route would follow the South Equatorial Current and the easterlies, leading them to the South Marquesas islands where the earliest evidence

---

[42] David Lewis, *We, the Navigators* […], on p. 274, 275.

[43] Magellan's route adapted from Tomás Mazón Serrano ("Mapas", Ruta Elcano, June 19, 2024, https://rutaelcano.com/mapas/); map adapted from Natural Earth (1:10 m large-scale land data, version 5.1.1, https://www.naturalearthdata.com/downloads/).

Fig. 15.7 Illustration of a Native American balsa raft, taken from Joris Van Spilbergen's report of the 1615 conquest of Paita, in modern-day Peru[46]

of admixture DNA was found. The 12th-century dating of this DNA is also consistent with the date the Polynesians settled these islands. If the replenished provisions included sweet potatoes—likely, since they were a staple food in Central America—this hypothetical journey could also explain its presence in Polynesia, and why its name—*kumala*—seems inspired by the Quenchuan *cumal*.[44] As enticing as this theory is though, it hinges on a long succession of fortuitous events.

This is nevertheless also the case with the alternative theory that the Native Americans were the ones sailing to the South Marquesas. These peoples did not seem to feel a strong inclination for oceanic exploration, and the meagre surviving records suggest their vessels and navigation techniques were less developed than those of the Polynesians. Still, there are some indications of long-distance maritime trade between present-day Ecuador and Mexico, a trip of about 3,100 km. This voyage was made with sizeable rafts—6 to 11 m long and capable of carrying 10 to 30 metric tonnes—with two sailing masts and a set of centreboards which were lowered into the water for steering and to counter the lateral forces of the sails[45] (Fig. 15.7).

[44] Alexander Ioannidis, Javier Blanco-Portillo et al., "Native American gene flow into Polynesia predating Easter Island settlement" [...], on p. 575.

[45] Leslie Dewan and Dorothy Hosler, "Ancient Maritime Trade on Balsa Rafts: An Engineering Analysis", *Journal of Anthropological Research*, 64(1), 2008), p. 19–40, on p. 26, 36.

[46] Anonymous, *Conquest of Paita: 1615* (Rijksmuseum, FMH 1055-AII(5)/13, 1646).

The Achilles heel of these rafts was their balsa wood construction. Balsa decomposes quickly when submerged in saltwater, due to its low density and the lack of chemicals to repel microorganisms and molluscs. Balsa is particularly vulnerable to shipworms, a voracious cellulose eating mollusc. Assuming shipworms were only later introduced in South America by Spanish ships, or alternatively that Native Americas coated their rafts with a protective layer of tar or beeswax (no archaeological evidence of such coatings was found so far, though), each raft could conceivably make two Ecuador-Mexico return trips before being scrapped. This means these balsa rafts could spend at least six months at sea before losing its buoyancy,[47] enough to reach the South Marquesas.[48] It is unlikely the crews of such vessels—which were engaged in local trade rather than deep-sea exploration—would embark willingly on such a voyage. Nevertheless, one could have been driven off their usual Ecuador to Mexico route by a storm, finding themselves drifting inexorably west under the Southern Equatorial Current and easterlies. This hypothetical journey would then explain the presence of Native American DNA in Polynesians, as well as the introduction of sweet potatoes.

Both these theories seem farfetched. On the one hand, the Polynesian return trip would require sailing about 10,000 km from Tahiti to Colombia; being aided by the Zenú natives; bringing a few aboard, together with sweet potatoes and other provisions; finding the easterlies and Southern Equatorial Current; and sailing another 7,000 km to the South Marquesas. On the other hand, the Native Americans would need to cover a shorter distance of about 7,000 km, but do so on a drifting and decomposing balsa raft. Once they arrived at the South Marquesas, they would need to be amicably received by the native Polynesians (or survive on their own until Polynesians arrived) and have some leftover sweet potatoes to plant.

Nevertheless, there is compelling evidence that one of these incredible voyages happened. And, after all, if not thoroughly documented, Magellan's expedition would appear equally improbable.

So then, who crossed the Pacific first? Magellan and his crew were most likely the first to ever cross the entire Pacific Ocean in the western direction, because both theories only take the Polynesians and Native Americans to the Southern Marquesas, less than half of the length of this ocean. The first ever Pacific crossing in the eastward direction is unclear. If Polynesians did indeed

---

[47] Leslie Dewan and Dorothy Hosler, "Ancient Maritime Trade on Balsa Rafts: An Engineering Analysis" […], on p. 36.

[48] The trip is about 7,000 km, meaning that the raft would need to make ~165 km per day, within reach of its estimated 7 km/h cruising speed (Leslie Dewan and Dorothy Hosler, "Ancient Maritime Trade on Balsa Rafts" […], on p. 36).

reach the Americas, then they were the first to do so, although in a piecemeal fashion. The first non-stop crossing was only accomplished several centuries later, in 1565, with the Spanish Manila to Acapulco route.

16

Around the World

From the Moluccas (December 21st, 1521) to Cape Roxo (June 28th, 1522)

On the 21st of December—four months before the *Trinidad* begun its fatidic last voyage—the *Victoria* raised anchors and left the Moluccas with a crew of 60 (47 Europeans and 13 Moluccans).[1] The crew of locals included a couple of pilots hired to guide the ship until Timor, where the *Victoria* would be able to gather additional supplies before venturing into the Indian Ocean (Fig. 16.1).[2]

By the 1520s, hundreds of Portuguese ships had already made the trip to the Indies. Starting with 1499's Vasco da Gama first expedition from Lisbon to Calicut, the Portuguese had turned the once unchartered waters into a sustainable trading route. Every year ships laden with spices left the Indies and made their way through a network of familiar ports: Malacca, Goa, Mozambique, Cape Verde and, finally, Lisbon. On each port, they found a trading post manned by friendly kinsmen who provided the required supplies and repairs. Even then, over 20% of the ships wrecked (Fig. 13.2).

The *Victoria* would of course have no access to such amenities. Worse still, courtesy of Lorosa,[3] they knew the Portuguese were actively searching for

---

[1] Antonio Pigafetta, *The first voyage around the world 1519–1522*, edited by Theodore Cachey Jr. (Toronto: University of Toronto Press, 2007), on p. 101; Medina, based on the available records, has listed the names of the 47 Europeans (José Toribio Medina, *El descubrimiento del Océano Pacífico: Vasco Núñez de Balboa, Fernando de Magallanes y sus compañeros,* Santiago de Chile, Imprenta Universitaria, 1920, on p. CCCX, CCCXI.

[2] José Toribio Medina, *El descubrimiento del Océano Pacífico: Vasco Núñez de Balboa, Fernando de Magallanes y sus compañeros* (Santiago de Chile: Imprenta Universitaria, 1920), on p. CCCXII.

[3] The disillusioned Portuguese trader they had met at the Moluccas, see The Moluccas.

R. Gaspar, *The Revolution of Magellan,* https://doi.org/10.1007/978-3-032-10797-8_16

Fig. 16.1 The *Victoria* route[4]

them. After the fleet left Spain, King Manuel I dispatched two fleets to intercept them, sending one to South America and one to the Cape of Good Hope (in case Magellan attempted to follow the Portuguese trading route). When both missions came back empty-handed, the king ordered Diogo Sequeira[5] to be on the lookout for them at the Moluccas and surrounding waters.[6]

So far they had managed to avoid the Portuguese patrols, but it would be foolish to push their luck. The plan was therefore to quickly leave the area and sail directly to Spain without ever raising land, a bold endeavour never achieved before. Elcano, now captain of the *Victoria*, would later proudly write to the Spanish king they had *"resolved to die rather than to fall into the hands of the Portuguese"*[7] (judging by what would happen to the *Trinidad's* crew,[8] the difference between the two options was slim).

A lot had to happen for Elcano—originally the *Concepcíon's* master— to find himself in the captain's chair of the *Victoria*. The first steps had taken place in the cold and faraway Patagonia with the killing of Juan de Elorriaga (the *San Antonio's* master), the execution of Gaspar de Quesada

---

[4] Fleet route adapted from Tomás Mazón Serrano ("Mapas", Ruta Elcano, June 19, 2024, https://rut aelcano.com/mapas/); map adapted from Natural Earth (1:10 m large scale land data, version 5.1.1, https://www.naturalearthdata.com/downloads/).

[5] A commander under whom Magellan had once served (see Winter in Patagonia and The Ambush), who was now responsible for all naval Portuguese forces in the Indies.

[6] José Toribio Medina, *El descubrimiento del Océano Pacífico* […], on p. CCCVII.

[7] "Carta de Juan Sebastián Elcano sobre su viaje de circunnavegación o primera vuelta al mundo" (Seville: Archivo General de Indias, ES.41091.AGI//PATRONATO,48,R.20, 1522), https://pares.mcu. es/ParesBusquedas20/catalogo/description/7343174.

[8] See The Eastern Dead-End.

(captain of the *Concepción*), and the marooning of Juan de Cartagena (*San Antonio's* captain). The desertion of the *San Antonio* took away additional higher-ranking officers: Álvaro de Mesquita (originally a supernumerary at the *Trinidad*, promoted to captain of the *San Antonio* by Magellan) and Estevão Gomes (originally the *Trinidad's* pilot, later transferred to the *San Antonio*). Antonio de Coca (the fleet's accountant) and Juan Rodríguez de Mafra (originally the *San Antonio's* pilot) succumbed to scurvy shortly after the fleet arrived at the Philippines. Magellan died at Mactan, together with Cristovão Rebelo, promoted to captain of the *Victoria's* shortly before. The ensuing ambush at Cebu took away Juan Rodríguez Serrano and Duarte Barbosa (who shared the chair of captain general after Magellan's death), and Andrés de San Martín (the fleet's astrologer-pilot). Finally, João Lopes Carvalho—the fleet's next captain general—was deposed by the crew and replaced by Elcano and Espinosa. Luck for the former and misfortune for the latter would put Elcano at the helm of the *Victoria*, leaving Espinosa in charge of the ill-fated *Trinidad*.

Elcano's ascension to power may have been driven by chance more than by merit, but the *Victoria's* new captain would soon prove worthy of his post by keeping the discouraged crew in check and adapting their original plan to the changing circumstances.

The first stages of the trip went by uneventfully. Guided by the two local pilots, the *Victoria* easily found its way first through the Molucca and Banda seas. However, in the Savu Sea a violent storm damaged the ship and threatened to send it crashing into the belt of northern islands which flank Timor. Fortunately, the crew was able to reach Alor Island, where they stayed for a couple of weeks performing repairs.[9] On January 25th they resumed the journey to Timor, the last planned stop before the dreaded oceanic crossing. As was his habit, Pigafetta merrily went ashore to lead the bartering efforts for fresh provisions. This time however, the locals demanded an exorbitant price for livestock, so the Europeans resorted to kidnapping a local chief and ransoming him for provisions.[10] Pigafetta recounted that the locals were afflicted by the "evil of St. Job" (syphilis), but it was probably leprosy.[11] Whatever it was it kept the crew away from the local women, preventing them from infuriating the natives.

---

[9] José Toribio Medina, *El descubrimiento del Océano Pacífico* [...], on p. CCCXII; Antonio Pigafetta, *The first voyage around the world 1519–1522*, edited by Theodore Cachey Jr. (Toronto: University of Toronto Press, 2007), on p. 114.

[10] Antonio Pigafetta, *The first voyage around the world 1519–1522* [...], on p. 117.

[11] Antonio Pigafetta, *The first voyage around the world 1519–1522* [...], on p. 178, 179.

The *Victoria* then continued its way down the northern coast of Timor. As they approached the western tip of the island and the prospect of a gruesome oceanic crossing drew closer, the crew became increasingly restless. An apprentice seaman and a cabin boy swam ashore and deserted, while several accounts mention a brawl or even a mutiny where several crewmen were killed.[12] Still, Elcano managed to keep the ship moving forward and, on February 8th, the *Victoria* sailed into the Indian Ocean. Aiming to cross the Cape of Good Hope well below the Portuguese shipping routes, they settled on a southwest course.

Before, while crossing the Pacific, the fleet missed many islands where they may have been able to replenish water and provisions.[13] This time they sighted Île Amsterdam, one of the very few stretches of land on the Southern Indian Ocean, but failed to find suitable anchorage.[14] Even if they had succeeded in coming ashore, it was unlikely they would find much there. The barren island, today part of the French overseas territories, is low on resources and hosts only Martin-de-Viviès, a small research facility.

The *Victoria* continued to follow a western course, but the trade winds made progress increasingly harder. Rations were down to water and rice, as the meat ransomed in Timor had spoiled for lack of salt to preserve it. The rundown ship also had a severe leak, forcing the hungry crew to spend hours at the bilge pumps. The physical exertion made perhaps the increasing cold slightly more bearable but did nothing for the empty stomachs. In April, facing the prospect of a mutiny, Elcano conceded to the pleas of his crew and ordered a change of course to the northwest. The plot was now to head towards Africa, hopefully finding a safe anchorage to reprovision while managing to slip through Portuguese ships.[15]

They reached the African coastline on May 8. Albo, leveraging the onboard nautical charts based on Portuguese surveys,[16] was able to identify the area as the mouth of the Infante River (today known as Great Fish River, in South Africa).[17] Against their best hopes, they were still east of the Cape of Good Hope and would spend the next ten days laboriously tacking against strong head winds to round the tip of Africa. Desperate for provisions, Elcano acquiesced to briefly stop at Saldanha Bay. The risky manoeuvre almost spelled the

---

[12] José Toribio Medina, *El descubrimiento del Océano Pacífico* [...], on p. CCCXIII.

[13] See Why Did Magellan Take Such a Convoluted Route Across the Pacific?

[14] Francisco Albo in Cristóbal Bernal, *Crónicas de la Primera Vuelta al Mundo, según sus Protagonistas,* 2016, on p. 411.

[15] José Toribio Medina, *El descubrimiento del Océano Pacífico* [...], on p. CCCXIV; Tim Joyner, *Magellan* (Maine: International Marine, 1992), on p. 229, 230.

[16] See How Did Magellan Convince the Spanish King?

[17] Francisco Albo in Cristóbal Bernal, *Crónicas de la Primera Vuelta al Mundo* [...], on p. 416.

end of the voyage, as they stumbled upon a Portuguese ship. As chance would have it, the Portuguese captain Pedro Quaresma was seemingly unaware of the naval feud with Spain or had more pressing matters to attend to, as he left the *Victoria* alone after only a curt exchange of words.[18] With the issue behind him, Elcano focused on sailing up the African coast, desperate to put an end to the exhaustion, malnutrition and scurvy steadily gnawing its way through the entire crew. They dared to approach land again on June 21st, near Cape Roxo (today the border of Guinea-Bissau and Senegal). Elcano sent a boat ashore to look for food and water and, while waiting for it to return, three more crewmen died. To the desperation of the survivors, the boat returned empty-handed.[19]

The situation was so dire it seems unlikely anybody aboard realised the momentous feat. Sometime around June 28th and in the vicinity of Cape Roxo they had crossed the path that, nearly three years ago, the fleet had taken on its way to Brazil. They had just circumvented the globe.

16.1 Who Were the First to Circumnavigate the Globe?

The short answer

The feat of circumnavigating the world for the first time has seemingly turned into the epitome for exploration, perhaps explaining why so many candidates for the role have been put forward. Some of the theories, such as those proposing Magellan or Enrique, are not supported by facts. Others are well-supported by evidence but do not fulfil all the requirements of what today is considered a true circumnavigation: starting and ending at the same point, crossing all longitudes, passing through at least a pair of antipodes, and travelling more than 40,000 km (the length of the equator). Elcano and his 17 companions who made it back to Seville on September 8th, 1522, are usually considered to be the first circumnavigators. However, this may be unfair. A few months before, in the vicinity of Cape Roxo, the Victoria crossed the path the fleet had taken on its way to Brazil. This moment also meets the criteria for a true circumnavigation, and nearly doubles the number of first circumnavigators from 18 to 32 (the Victoria's crew at that moment).

Were Elcano and his surviving crew the first to ever complete a circumnavigation? The topic is somewhat contentious, with at least six different theories often brought forward (Fig. 16.2).

---

[18] José Toribio Medina, *El descubrimiento del Océano Pacífico* [...], on p. CCCXIV, CCCXV.
[19] Tim Joyner, *Magellan* [...], on p. 231.

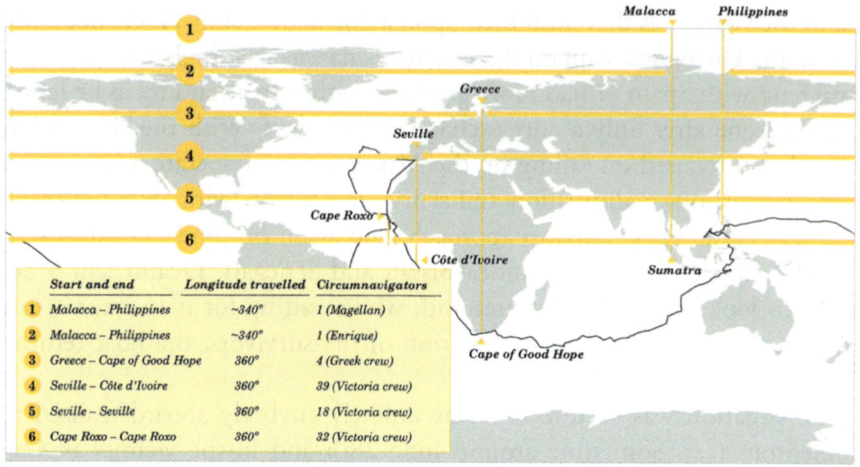

	Start and end	Longitude travelled	Circumnavigators
1	Malacca – Philippines	~340°	1 (Magellan)
2	Malacca – Philippines	~340°	1 (Enrique)
3	Greece – Cape of Good Hope	360°	4 (Greek crew)
4	Seville – Côte d'Ivoire	360°	39 (Victoria crew)
5	Seville – Seville	360°	18 (Victoria crew)
6	Cape Roxo – Cape Roxo	360°	32 (Victoria crew)

Fig. 16.2 Overview of six popular candidates for the world's first circumnavigation

In the first theory,[20] Magellan was the world's first circumnavigator, albeit in a piecemeal fashion: first, by travelling to the longitude of the Moluccas while serving the Portuguese crown,[21] and then by reaching Mactan, in the Philippines (the Moluccas are 3° east of Mactan). While same sources insinuate Magellan could have visited the Banda Islands (just south of the Moluccas),[22] most records—including the official Portuguese rosters—never place Magellan east of Malacca,[23] suggesting he fell about 20° short of completing a full circumnavigation before dying at Mactan.

The second hypothesis[24] makes Enrique—Magellan's slave and the fleet's interpreter—the first to circle the Earth, when the fleet reached Cebu, in the Philippines. For the claim to hold, Enrique could not have been from Malacca (as stated by Magellan[25]) or Sumatra (as reported by Pigafetta[26])— but instead from the Philippines. While it is possible he was abducted from a

---

[20] Suggested, for instance, by José Manuel Garcia (*Fernão de Magalhães: Herói, traidor ou mito*, Lisboa, Editorial Presença, 2020).

[21] See Why Was Magellan, a Portuguese Citizen, Serving Spain?

[22] José Manuel Garcia, "Fernão de Magalhães o homem que descobriu o mundo tal como ele é", *Revista Mátria XXI*, 2021, p. 313–327, on p. 314–326.

[23] Jean Denucé, *Magellan: la question des Moluques et la première circumnavigation du globe* (Brussels: Hayez, 1911), on p. 121, citing Góis, Castanheda and Correia.

[24] Suggested, for instance, by Penélope Flores ("Magellan's Interpreter, Enrique, Was the First to Circumnavigate the World", Positively Filipino, retrieved November 2024, https://www.positivelyfilip ino.com/magazine/magellans-interpreter-enrique-was-the-first-to-circumnavigate-the-world).

[25] Magellan's will in Francis H. H. Guillemard, *The life of Ferdinand Magellan and the first circumnavigation of the globe: 1480–1521* (New York: Dodd Mead, 1890), on p. 321.

[26] Antonio Pigafetta, *The first voyage around the world 1519–1522,* edited by *Theodore Cachey Jr.* (Toronto: University of Toronto Press, 2007), on p. 34.

village there and sold as a slave in Malacca, where Magellan bought him, the only evidence supporting such a claim is that Enrique was able to communicate with the Filipinos. However, rather than communicating in Visayan—the dialect likely spoken in Central Philippines at the time—he may have used Malay, the region's lingua franca.[27] Such an explanation would be more consistent with Pigafetta's account of what happened at Limasawa, as he stated that only the king but not the natives understood Enrique.[28] Alternatively, Enrique could have travelled back to Sumatra after deserting the fleet at Cebu.[29] Assuming he was not killed or resold as a slave along with the Europeans, it is indeed feasible that he found his way back home, but there is no evidence he ever did. It seems therefore more likely that Enrique, as Magellan, fell ~20° short of a circumnavigation.

In the third theory,[30] there was not one but four circumnavigators: Francisco Albo, Miguel de Rodas, Nicolás el Griego, and Miguel Sanchéz. These mariners came from Greece, and therefore completed a 360° journey when the *Victoria* was close to the Cape of Good Hope, which sits at approximately at the same longitude (Fig. 16.2). Unlike the previous two hypothesis, this one is consistent with all the available evidence, but maybe one needs to establish some ground rules on what exactly constitutes a circumnavigation. After all, someone at the South Pole can make such a claim only by taking a stroll around the pole. For the Guinness World Records, a true circumnavigation starts and ends at the same point, crosses all longitudes, passes through at least a pair of antipodes,[31] and takes more than 40,000 km (the length of the equator). Unfortunately for the four Greeks, their journey did not start and end at the same point.

Theory number four,[32] which recognizes 39 circumnavigators (the *Victoria's* crew members still alive when the ship returned to the longitude of Seville, while south of Cote d'Ivoire), also fails for the same reason.

---

[27] Robert Kaplan and Richard Baldauf Jr., *Language and Language-in-Education Planning in the Pacific Basin* (Dordrecht: Kluwer Academic Publishers, 2003), on p. 83.

[28] "*When they were near the ship of the captain-general, the said slave spoke to the king, who understood him well, because in these countries the kings know more languages than the common people*" (Antonio Pigafetta in H.E.J. Stanley, *The first voyage round the world* […], on p. 76).

[29] See The Ambush.

[30] Suggested (and rejected) by Miguel Zafra Caramé ("Los Primeros Circunnavegantes", *Revista General de Marina*, January-Februeary 2022, p. 5–14, on p. 13, 14).

[31] Two points in the surface of the Earth that are diametrically opposite to one another.

[32] Suggested, for instance, by Miguel Zafra Caramé, "Los Primeros Circunnavegantes" […], on p. 11–13.

Instead, it was Elcano and the remaining 17 sailors who returned to Seville on the 8th of September 1522[33] who made it into the coveted book of Guinness World Records.[34] Indeed, this fifth theory ticks all rules: it started and ended in the same point (Seville), crossed all longitudes, passed through at least a pair of antipodes (for instance, Guam and Salvador, which the fleet sailed past on their way to Rio de Janeiro), and sailed more than 40,000 km (about 69,000 km[35]). Guinness was not the first proponent of such a hypothesis, as the Spanish king ennobled Elcano with a coat of arms with the words *Primus circumdedisti me* ("You were the first to encircle me") wrapped around a globe.

There is however a sixth and final theory[36] which recognizes 32 circumnavigators, the *Victoria*'s crew when the ship crossed paths with the expedition's outgoing route, somewhere near Cape Roxo (Fig. 16.2). This proposal also fulfils all of Guinness' requirements for a true circumnavigation, assuming the start and end point of the journey can be located on water. That seems to be a reasonable assumption, so Table 16.1 recognizes these 32 mariners. Not all of them accompanied Elcano in his triumphal return to Spain (two died and 12 lingered in a Cape Verdean jail[37]), but all arguably deserve to be part of the Guinness World Records.

16.2 Why Did Nautical Charts Ignore the Earth Was Round?

The short answer

For some 500 years, from the late 13th century to well into the 18th century, ships navigated by nautical charts that lacked a projection, i.e., which assumed the Earth was flat. Of course, none of the chart makers and pilots were flat earthers. Instead, they were bound by the limitations of early modern science. In 1569, Gerardus Mercator created the first projection useful for navigation (a conformal projection, showing courses of constant heading as straight lines). Before it could be used though, three additional breakthroughs were necessary: surveys to accurately determine the longitude of places, mappings of the spatial distribution of magnetic declination, and a dependable way to measure longitude on board.

---

[33] See The Lucky Seven Percent Get Home.

[34] "First Circumnavigation", Guinness World Records, assessed November 2024, https://www.guinnessworldrecords.com/world-records/first-circumnavigation.

[35] Tomás Mazón Serrano, "La primera vuelta al mundo", Ruta Elcano, assessed November 2024, https://rutaelcano.com/la-primera-vuelta-al-mundo-parte-ii/.

[36] Suggested, for instance, by Miguel Zafra Caramé ("Los Primeros Circunnavegantes" […], on p. 9–11).

[37] See The Lucky Seven Percent Get Home.

Table 16.1 Name, role, and fate of the 32 sailors who were part of the *Victoria's* crew when she completed a circumnavigation near Cape Roxo[38]

Name	Role	Outcome
Esteban Villón	Able seaman	Died (August 6th, 1522)
Andrés Blanco	Apprentice seaman	Died (July 14th, 1522)
Martín Méndez	Clerk	Imprisoned at Cape Verde
Ricarte de Normandía	Carpenter	Imprisoned at Cape Verde
Roldán de Argote	Gunner	Imprisoned at Cape Verde
Felipe de Rodas	Able seaman	Imprisoned at Cape Verde
Gómez Hernández	Able seaman	Imprisoned at Cape Verde
Ocacio Alfonso	Able seaman	Imprisoned at Cape Verde
Pedro de Tolosa	Apprentice seaman	Imprisoned at Cape Verde
Pedro de Churdurza	Cabin boy	Imprisoned at Cape Verde
Vasquito Gallego	Cabin boy	Imprisoned at Cape Verde
Maestre Pedro	Supernumerary	Imprisoned at Cape Verde
Juan Martin	Supernumerary	Imprisoned at Cape Verde
Simón de Burgos	Supernumerary	Imprisoned at Cape Verde
Juan Sebastián de Elcano	Captain	Returned to Seville (Sep 8th, 1522)
Francisco Albo	Pilot	Returned to Seville (Sep 8th, 1522)
Miguel de Rodas	Master	Returned to Seville (Sep 8th, 1522)
Juan de Acurio	Master's mate	Returned to Seville (Sep 8th, 1522)
Hernando de Bustamante	Barber	Returned to Seville (Sep 8th, 1522)
Maestre Hans	Gunner	Returned to Seville (Sep 8th, 1522)
Martín de Judícibus	Man-at-arms	Returned to Seville (Sep 8th, 1522)
Diego Carmena Gallego	Able seaman	Returned to Seville (Sep 8th, 1522)
Nicolás el Griego	Able seaman	Returned to Seville (Sep 8th, 1522)
Miguel Sánchez	Able seaman	Returned to Seville (Sep 8th, 1522)
Francisco Rodríguez	Able seaman	Returned to Seville (Sep 8th, 1522)
Juan Rodríguez	Able seaman	Returned to Seville (Sep 8th, 1522)
Antón Colmenero	Able seaman	Returned to Seville (Sep 8th, 1522)
Juan de Arratia	Apprentice seaman	Returned to Seville (Sep 8th, 1522)

(continued)

---

[38] Adapted from Miguel Zafra Caramé, "Los Primeros Circunnavegantes" [...], on p. 17.

Table 16.1   (continued)

Name	Role	Outcome
Juan de Santandrés	Apprentice seaman	Returned to Seville (Sep 8th, 1522)
Vasco Gómez Gallego	Apprentice seaman	Returned to Seville (Sep 8th, 1522)
Juan de Zubileta	Cabin boy	Returned to Seville (Sep 8th, 1522)
Antonio Pigafetta	Supernumerary	Returned to Seville (Sep 8th, 1522)

Not even the most uneducated sailor aboard Magellan's fleet believed the Earth was flat. Why would they embark on a journey to reach the East by travelling west if they did? However, they were assuredly not aware the charts guiding their voyage still assumed the Earth to be flat.

More precisely, these charts lacked a projection, i.e., a way to correctly represent a spherical surface on a plane. Anybody who tries to lay an orange peel flat on a surface will quickly realise it cannot be done. Without the ability to stretch the peel in some places and contract it in others, there will always be gaps. Maps of any sort—including nautical charts—suffer from the same issue: without a projection to mathematically define which areas to contract and to expand, a consistent representation of the surface of the Earth cannot be presented on a piece of paper.

The first latitude charts used for oceanic crossings were based on latitude, heading and, as a complement, distance estimates.[39] These measurements were then transferred directly from the surface of the Earth to the chart's plane without using a projection, as if the ships travelled in a straight line rather than in an arc following the curvature of the Earth.[40] How were the world's best chart makers capable of such an oversight, given that projections existed since ancient Greece[41]? And how could these charts be useful at all?

---

[39] See How Did Magellan Convince the Spanish King?

[40] Joaquim Gaspar, "The Myth of the Square Chart", *e-Perimetron*, 2(2), 2007, p. 66–79, on p. 78.

[41] At least since Marinus of Tyre (c.70–c.130 AD) (Mark Monmonier, *Rhumb lines and map wars: A social history of the Mercator projection*, Chicago, The University of Chicago Press, 2004, on p. 28).

The projection-less nature of latitude charts was inherited from Mediterranean portolan charts. When these were created, probably in the early thirteenth century,[42] many ancient Greek findings had been long forgotten in Europe, including the projection that Marinus of Tyre had proposed more than a millennia before. In the early fifteenth century, when these texts were translated from Greek to Latin and re-introduced in Europe, neither the chart makers building portolan charts nor the pilots using them were in any hurry to use projections. After all, why would they add additional complexity to something which was accurate enough to navigate the confined and well-known Mediterranean waters?

By the early sixteenth century, portolan charts gave way to latitude charts, which were required to navigate through vast expanses of water. A hundred years had passed since the reintroduction of Marinus of Tyre projection, but it continued to be ignored by chart makers and pilots, despite the considerable authority that ancient knowledge commanded in mediaeval Europe. Surely there was a heavy price to pay for such an oversight, with many ships and lives lost to the perils of projection-less navigation? Not at all. In fact, using these projections would have made latitude charts less precise, not more.

Marinus of Tyre had proposed an equidistant cylindrical projection.[43] One way to visualise it is to first draw a square graticule onto a piece of paper, turn it into a cylinder, and wrap it around a world globe. Imagine now that the surface of the globe is made of malleable rubber. Peel it and stick it onto the cylinder, in such a way its meridians and parallels align with the cylinder's square graticule. This is straightforward to do near the Equator, where the cylinder and the globe touch. As one gets closer to the poles, the surface of the globe needs to be more and more stretched horizontally to fit the surface of the cylinder without any gaps. This means a circle placed on the surface of the globe on the Equator is still a circle on the projection but becomes increasingly elongated when it approaches the poles. In the poles, the circle turns into a straight segment (Fig. 16.3).

Marinus of Tyre's equidistant cylindrical projection is not the only one which causes distortions. *Any* projection will distort the shapes of land masses. As illustrated by the orange peel experiment, there is no way to transpose the surface of a sphere to a plane without some degree of stretching and squeezing. Distortion is minimal around the standard parallel—the tangent

---

[42] Joaquim Gaspar, "The origin of nautical cartography: certitudes, doubts, and perplexities", *International Journal of Cartography*, 2023, p. 1–30, doi:10.1080/23729333.2023.2240902, on p. 6.
[43] Mark Monmonier, *Rhumb lines and map wars* […], on p. 28.

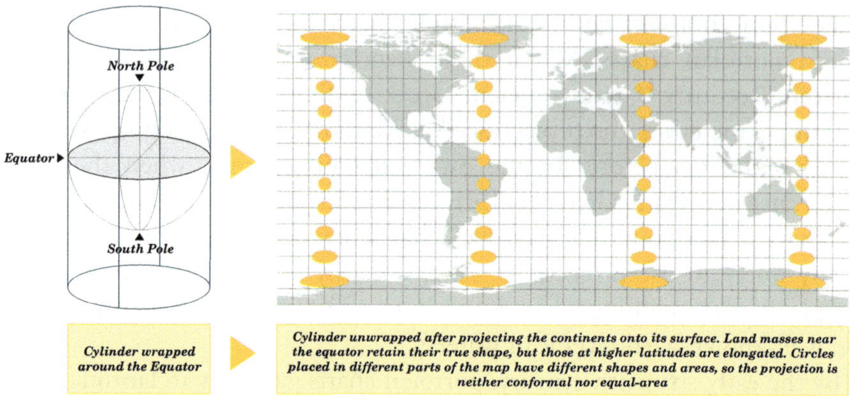

Fig. 16.3 Map distortions in an equidistant cylindrical projection, as proposed by Marinus of Tyre[44]

where the globe meets the projection's plane—and increases with the distance to that standard parallel.

While it is not possible to conserve shapes, it is possible to control the stretching and squeezing to achieve one, and only one, of the following: to conserve areas, or to conserve angles (Fig. 16.4). A projection which conserves areas is called *equal-area* and is useful when comparing relative sizes is important—for instance, to show Greenland is in fact much smaller than Africa. A projection that conserves angles is called *conformal* and is useful for navigation, as routes of constant heading are represented as straight lines.[45] This means a pilot going from point A to Point B can take a conformal chart, draw a straight line between the two points, measure the angle of the line, and then use a compass to follow that constant heading once underway. Marinus of Tyre's equidistant cylindrical projection was not conformal,[46] so it was not good for navigation.

---

[44] WGS 84 / Plate Carrée (ESPG:32,662) projection; map adapted from Natural Earth (1:10 m large scale land data, version 5.1.1, https://www.naturalearthdata.com/downloads/); Tissot indicatrix adapted form Ervin Wirth and Péter Kun (Indicatrix Mapper v.2.0.2, https://github.com/ervinwirth/indicatrix-mapper).

[45] For additional information on the properties of map projections and guidance on which one to choose for a given application, see John Snyder and Philip Voxland, *An Album of Map Projections* (Washington D.C., United States Government Printing Office, 1989).

[46] Nor was it equal-area. The first equal-area projection dates from the late eighteenth century, and was created by Johann Heinrich Lambert, a Swiss mathematician (John Snyder, "The Transverse and Oblique Cylindrical Equal-Area Projection of the Ellipsoid", *Annals of the Association of American Geographers*, 75(3), 1985, p. 431–442, on p. 431).

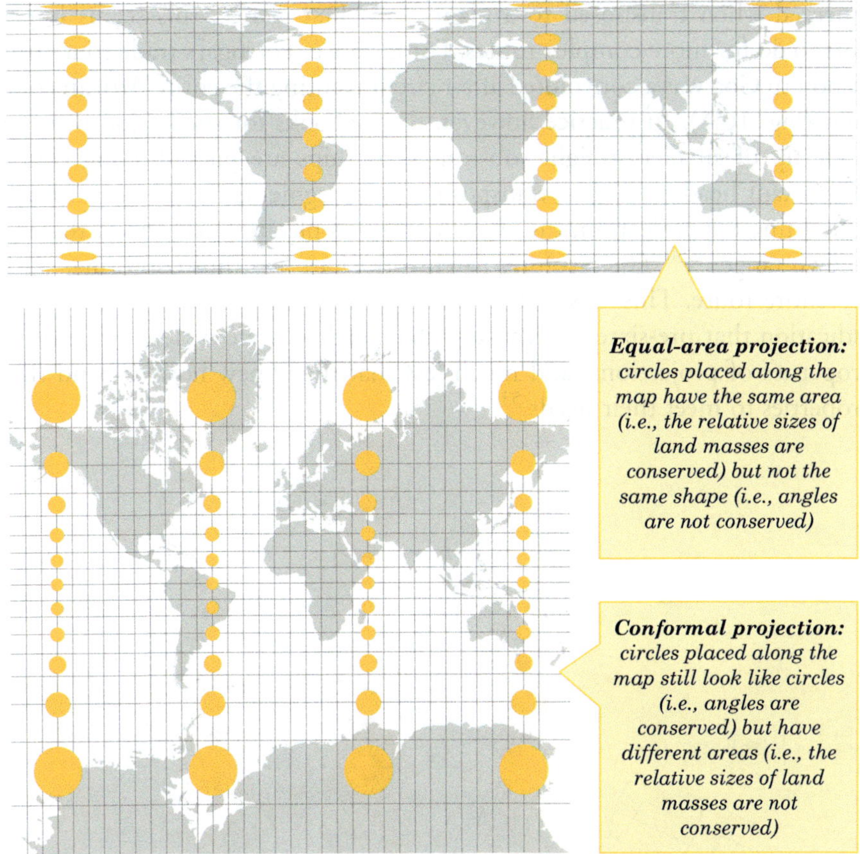

Fig. 16.4 Map distortions in equal-area and conformal projections[47]

The odd thing is that portolan and latitude charts, despite lacking a projection altogether, did conserve (some) angles.[48] How could a projection-less chart be (somewhat) conformal?

By transposing pilot readings directly to the chart without accounting for projection, chart makers were implicitly assuming the route had been made along the tangent where the globe meets the chart's plane, which is the only place where the surface of the sphere does not need to be stretched nor

---

[47] Cylindrical Equal Area (ESRI:54,034) and WGS 84 / World Mercator (ESPG:3395) projections; map adapted from Natural Earth [...]; Tissot indicatrix adapted form Ervin Wirth and Péter Kun [...].

[48] Mark Monmonier, *Rhumb lines and map wars* [...], on p. 23.

squeezed when transposed to a plane.[49] At first glance, this seems impossible to do for more complex routes. For instance, Portugal's trade route from Lisbon to Goa was a convoluted affair circumventing the entire African continent (Fig. 16.5). In practice however, the logbooks for such a journey were made up of smaller straight sub-routes, as ships are piloted along courses of constant heading. When transposing these segments to the chart without accounting for projection, the chart maker was placing each one of them on its own tangent.[50] The resulting patchwork chart was thus conformal across the entire route. This was fortuitous rather than intentional, as there is no indication that pre-sixteenth century chart makers were widely aware of the properties of projections, and much less that they knew how to bend these properties to meet their needs.[51]

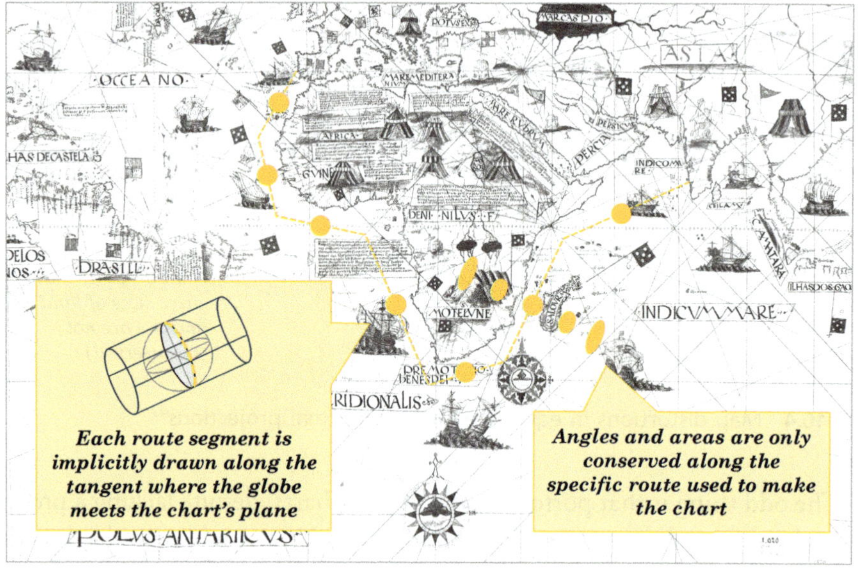

Fig. 16.5 Example of how a nautical chart based on the Lisbon-Goa trading route maintained conformity along that track, but was distorted elsewhere[52]

---

[49] Idem on p. 28; Joaquim Gaspar and Henrique Leitão, "What is a nautical chart, really? Uncovering the geometry of early modern nautical charts", *Journal of Cultural Heritage*, 29, 2018, p. 130–136, on p. 132, 133.

[50] In other words, instead of having a single standard parallel, as an equidistant cylindrical projection (Fig. 16.3), latitude charts have multiple "standard lines", one for each route segment.

[51] Joaquim Gaspar and Henrique Leitão, "What is a nautical chart, really?" [...], on p. 133.

[52] Tissot indicatrix are illustrative only; using the Kunstmann IV chart as an example (attributed to Jorge Reinel and Pedro Reinel, *Kunstmann IV Nautical Planisphere*, facsimile by Otto Progel, Paris, Bibliothèque Nationale de France, CPL GE AA-564, c1519).

There was however a hefty price to pay for ignoring the laws of projection. Latitude charts were only conformal along the specific routes which had been used to build them, and distortions increased with the distance to those routes. And, as seen before,[53] these charts were also warped by the effects of magnetic declination. Using them on novel routes meant sending ships and sailors into the projection-less and distorted unknown.

Latitude charts were not so much maps as they were breadcrumb paths which guided ships along familiar routes. The charts aboard Magellan's ships were accurate for navigating along the Atlantic and Indian Ocean routes that have been used to build them, but they were untrustworthy elsewhere.

The first conformal projection was discovered by a Flemish geographer named Gerardus Mercator in 1569. A couple of decades before, in 1537, a Portuguese mathematician called Pedro Nunes had realised a ship maintaining a constant heading would not travel along a great circle—the shortest distance between two points on the surface of a sphere—but would instead follow a spiral path, today known as a loxodrome or rhumb line (Fig. 16.6).[54] To create his projection, Mercator progressively increased the spacing of parallels from the Equator to the poles, turning this loxodrome into a straight path.[55]

The Flemish geographer showcased his novel projection in a nautical chart called *"A new and augmented description of Earth corrected for the use of sailors"*.[56] Indeed, Mercator's projection held the potential to transform navigation, moving it away from the limitations of the latitude chart. Unfortunately, it was much ahead of its time. The issue with Mercator's projection was not its mathematical foundation, which was sound, but rather its geographical errors. In other words, Mercator's graticule of meridians and parallels was correct, but the superimposed land contours and coordinates of name places were wrong.[57]

---

[53] See How Did Magellan Convince the Spanish King?

[54] W. G. L. Randles, "Pedro Nunes' Discovery of the Loxodromic Curve (1537)", *The Journal of Navigation*, 50(1), 1997, p. 85–96, doi:10.1017/S0373463300023614, on p. 88.

[55] Mercator never described how he built his projection, but the most likely hypothesis is that he charted tables of rhumbs (sets of latitude and longitude coordinates describing several loxodromes) made by someone else (Henrique Leitão and Joaquim Gaspar, "Globes, Rhumb Tables, and the Pre-History of the Mercator Projection", *Imago Mundi*, 66(2), 2014, p. 180–195, doi: 10.1080/030 85694.2014.902580, on p. 180).

[56] Gerardus Mercator, *Nova et Aucta Orbis Terrae Descriptio ad Usum Navigantium Emendata* (France: Bibliothèque Nationale de France, GE A-1064 RES, 1569).

[57] Joaquim Gaspar, "Revisiting the Mercator World Map of 1569: an Assessment of Navigational Accuracy". *Journal of Navigation*, 69(6), 2016, p. 1183–1196, doi: 10.1017/S0373463316000291, on p. 1188–1194.

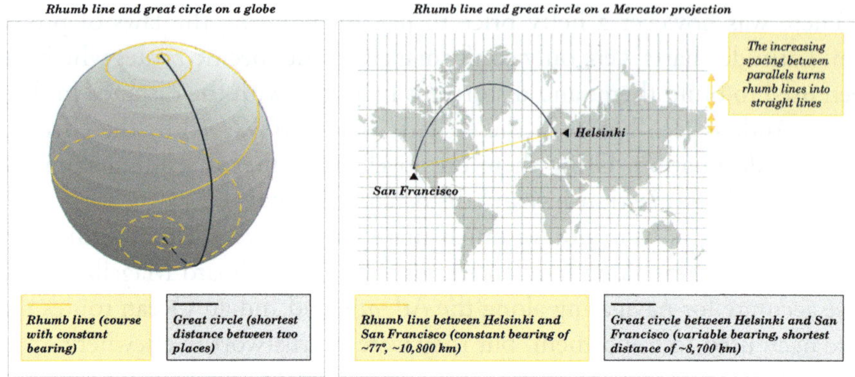

Fig. 16.6 Left: a rhumb line and a great circle plotted onto the surface of a globe; Right: a rhumb line and a great circle plotted onto a conformal map, illustrating how the former appears as a straight line[58]

Some of these errors were merely the result of the lack of information at the time: for instance, Europeans did not yet know about the existence of Australia and New Zealand, and the shape of the Americas was still very much speculative. Others—such as the stretched shape of Africa—were the result of more subtle issues lurking in the background (Fig. 16.7).

Mercator based his map on Portuguese and Spanish latitude charts without understanding how these charts had been built. Like most of his contemporaries, Mercator assumed they were based on an equirectangular projection, the same Marinus of Tyre had proposed almost 1,500 years before. So, he simply took the latitude and apparent longitude coordinates of each name place and transposed it to his new projection. In doing so, he carried over the distortions of latitude charts caused by the lack of projection and magnetic declination.[59]

Even if Mercator had realised the limitations of latitude charts, he would have had no other choice than to use them, as there were no other sources available.[60] Mercator did the best he could but still fell short of creating a viable alternative for navigation. It would take more than 200 years for his projection to replace latitude charts.

During those 200 years, three things happened: one, the longitude of places was measured, slowly replacing the estimated longitudes from latitude charts; two, the spatial distribution of magnetic declination was mapped, allowing for the use of corrected compass directions; and three, an accurate

---

[58] Example using a Google Maps Global Mercator (EPSG:900,913) projection.

[59] Joaquim Gaspar, "Revisiting the Mercator World Map of 1569" [...], on p. 1188–1194.

[60] Joaquim Gaspar, "Revisiting the Mercator World Map of 1569" [...], on p. 1194.

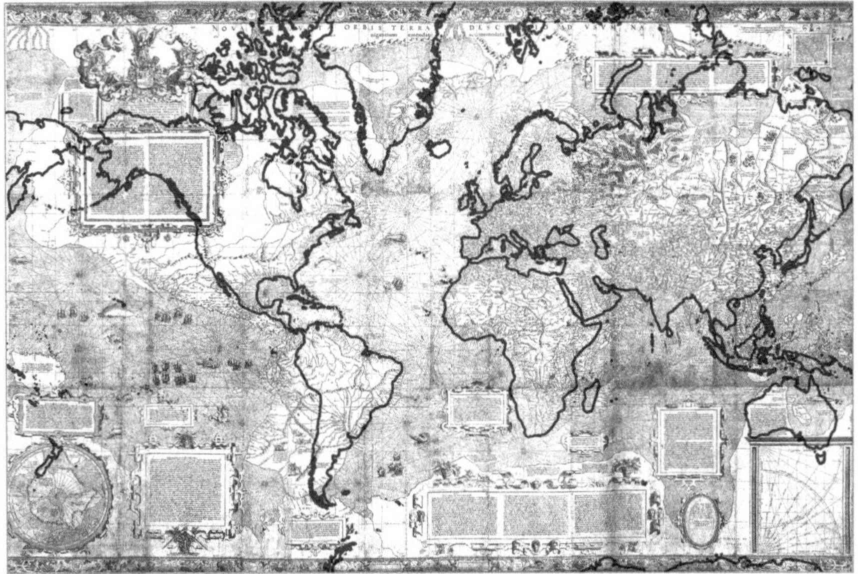

Fig. 16.7 Mercator's nautical chart overlaid with an accurate representation of the continents[61]

method for measuring longitude on board was developed, allowing pilots to finally compare the ship's position with the coordinates provided by a Mercator projection. It was only then that pilots started using latitude and longitude—rather than latitude and heading—to plot courses and track the ship's progress.[62]

For some 500 years, from the late thirteenth century to well into the eighteenth century, navigation was mostly based on projection-less charts and uncorrected compasses. Despite these scientific and technical limitations, the planet went from being a collection of isolated land masses to a grid of civilizations interconnected by global trade.

---

[61] Gerardus Mercator, *Nova et Aucta Orbis Terrae Descriptio ad Usum Navigantium Emendata* […]; accurate representation from continents adapted from Natural Earth (1:10 m large scale land data, version 5.1.1, https://www.naturalearthdata.com/downloads/).

[62] Joaquim Gaspar, "Revisiting the Mercator World Map of 1569" […], on p. 1195.

17

The Lucky Seven Percent Get Home
From Cape Roxo (June 28th, 1522) to Seville (September 8th, 1522)

Unable to find provisions in Cape Roxo and with just a few men still able to operate the bilge pumps keeping the ship afloat, the officers took a vote and agreed to head to the nearby Cape Verde islands,[1] a Portuguese territory (Fig. 17.1). To avoid being arrested on the spot, they would try to convince the port officials they were coming back from the Americas when a storm damaged their ship and separated them from the rest of the fleet .[2]

Elcano sent ashore the ship's clerk, the master-at-arms, and "Manuel", one of the surviving natives from the Moluccas (at this point, only four of the initial 13 Moluccans remained). The ruse seems to have been well-rehearsed, as the ship was authorized to stay and reprovision. Over the next days the *Victoria* received a few provisions, but not yet enough for the trip back to Spain.[3]

In the meantime, Pigafetta kept his mind busy with something that had been bothering him since they have arrived at Cape Verde: the port officials mentioned it was Thursday but, according to his records, it was Wednesday. He had flipped through this journal, where he meticulously kept track of dates, and found no missing entries.[4] Both Pigafetta and the port officials

---

[1] Francisco Albo in Cristóbal Bernal, *Crónicas de la Primera Vuelta al Mundo, según sus Protagonistas,* 2016, on p. 423, 424.

[2] Antonio Pigafetta, *The first voyage around the world 1519–1522,* edited by Theodore Cachey Jr. (Toronto: University of Toronto Press, 2007), on p. 125.

[3] José Toribio Medina, *El descubrimiento del Océano Pacífico: Vasco Núñez de Balboa, Fernando de Magallanes y sus compañeros* (Santiago de Chile: Imprenta Universitaria, 1920), on p. CCCXV–CCCXVII, citing Albo, Pigafetta, and Simón de Burgos (a supernumerary).

[4] Antonio Pigafetta, *The first voyage around the world* […], on p. 125.

© The Author(s), under exclusive license to Springer Nature Switzerland AG 2026

R. Gaspar, *The Revolution of Magellan,* https://doi.org/10.1007/978-3-032-10797-8_17

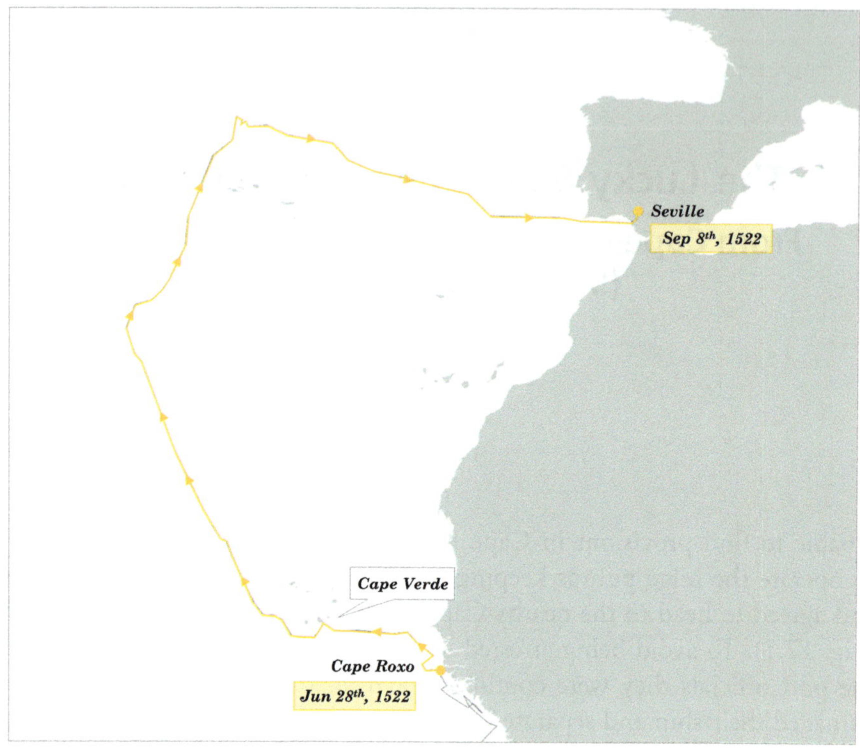

Fig. 17.1 The *Victoria* route[5]

were right: on the island's local time it was already Thursday, but on the ship's local time it was still Wednesday.

Pigafetta was the first to experience in the flesh the circumnavigator's paradox, described in the thirteenth century by Abulfeda (1273–1331), a Syrian geographer and historian. Abulfeda first assumed it was possible to complete a journey around the world—not something to take for granted in the thirteenth century. He then pictured three travellers. The first one went west, the second one east, and the third one stood still. After seven days the man sitting still saw both of his companions returning. The one travelling west reported his voyage had taken only six days, while the one travelling east reported he had needed eight days to cover the same distance.[6]

While the journey had taken the same time for both travellers, the issue was they were using the daily movement of the Sun to count the passage of

---

[5] Fleet route adapted from Tomás Mazón Serrano ("Mapas", Ruta Elcano, June 19, 2024, https://rut aelcano.com/mapas/); map adapted from Natural Earth (1:10 m large scale land data, version 5.1.1, https://www.naturalearthdata.com/downloads/).

[6] Geoffrey Gunn, *Overcoming Ptolemy: The Revelation of an Asian World Region* (Lanham: Lexington Books, 2018), on p. 47, 48.

Fig. 17.2 Current configuration of the International Date Line[9]

time. For the traveller going east, each day was one-seventh shorter, so he saw the sun setting eight times. Conversely, for *the traveller going west each day one-seventh longer, so he only saw the sun setting six times*. Some 200 years after, the *Victoria's* crew had proved Abulfeda right, albeit requiring more than a thousand sunsets, rather than six.

Much like how a two-dimensional map requires a projection to accurately depict a curved surface,[7] using the planet's revolutions to count the passage of time requires a time discontinuity somewhere. By 1565, when Spain occupied modern-day Philippines,[8] the nuisance of such a discontinuity became apparent, as the Philippines were one day ahead of the other Spanish territories. To have the same date across the entire Spanish Empire, the Filipinos saw their calendar moved back. They were kept one day behind all its Asian neighbours until 1884, when the International Date Line (IDL) was established, 180° from Greenwich (Fig. 17.2).

The IDL is not straight, as some zigs and zags are necessary to avoid blindly cutting across regions (e.g., the Russian Far East). Neither is it set

---

[7] See Why Did Nautical Charts Ignore the Earth Was Round?

[8] See What Happened to the Philippines?

[9] International Date Line from Google Maps (https://www.google.com/maps, retrieved December 2024); map adapted from Natural Earth [...].

in stone. Samoa hopped over the line twice, one in 1892 (to facilitate trade with the Americas) and again in 2011 (rejoining Australia and New Zealand, by then its major trading partners). Kiribati is responsible for the IDL's hammer-looking indent, when it shifted all its islands to the eastern side in 1994.

Back at Cape Verde, the circumnavigator's paradox was the least of their worries. Still low on supplies, Elcano sent the ship's longboat ashore with a crew of 13 to source them. The first trip went smoothly, and the longboat came back with a supply of rice. Still needing more provisions and slaves to operate the pumps, Elcano sent the longboat back to shore. This time, it did not return. Worried the port officials had caught word the supplies had been paid with cloves—the only thing of value they had to trade—Elcano ordered the *Victoria* to cautiously approach the harbour to see what had happened to his crew. As they approached the shore, the port officials hailed the vessel and ordered it to surrender. In no condition to fight they fled, leaving the 13 men behind.[10]

Elcano probably assumed it had been the use of cloves which had exposed their ploy, but it might not have been the case. Among the men sent ashore was a servant called Simão de Burgos. Burgos was Portuguese but, like other members of Magellan's expedition, had concealed his true nationality to bypass the quotas on Portuguese mariners imposed by the Spanish king. Impelled perhaps by a sense of patriotic duty or in hopes of gaining an advantage, Burgos may have revealed to the port officials the *Victoria* had not come from the Americas after all and was instead the surviving ship of Magellan's expedition.[11]

Back in open waters, the provisions they had sourced brought some energy back to the dying crew. However, down 13 men and with the ship badly leaking, the trip back to Spain was everything but easy. After nearly three months of sailing through uncooperative winds while desperately manning the bilge pumps, the *Victoria* finally limped back to Sanlúcar de Barrameda. Following a hearty meal of bread, meat, fruit and much missed wine, a local pilot and a six-oared longboat towed the battered ship up the Guadalquivir River. On September 8th 1522, nearly three years after it had left, the *Victoria* was back at Seville. The original fleet of five ships and the crew of 265 had

---

¹⁰ José Toribio Medina, *El descubrimiento del Océano Pacífico* [...], on p. CCCXVI, CCCXVII; the 13 men left behind include 12 Europeans (Table 16:1) and Manuel, the Moluccan.

¹¹ José Toribio Medina, *El descubrimiento del Océano Pacífico* [...], on p. CCCXVII–CCCXIX.

been reduced to the lonely *Victoria* with 18 haggard sailors aboard, plus three Moluccans.[12]

17.1 How Did Magellan's Expedition Change the World?

The short answer

For Magellan and most of his crew, the expedition was an unmitigated failure, as the captain and two thirds of the men perished. For Spain, its immediate effects were lacklustre, since the sale of the cloves brought back by the Victoria was not enough to cover the fleet's costs, and the position of the Moluccas relative to the Tordesillas anti-meridian remained unclear. However, the scales quickly tipped in 1529, when Spain sold its claims over the islands for more than 13 times what Magellan's expedition had cost.

Yet, the most valued thing brought back by the Victoria were not cloves but the knowledge that all the globe's major oceans were interconnected and navigable, paving the way for the hegemony of the ship. An unparalleled European expansion followed, one which only peaked 400 years later with the British Empire engulfing a quarter of the planet. The strategy was the same almost everywhere: first take over the major trading routes and ports, and then expand inland. The effects of European colonialism are visible in many of today's borders, cultures, and religions. However, nowhere on Earth were its effects as marked as in the New World, where some 70% of the Native American population died after the arrival of Europeans, giving way to more than 12 million slaves brought from Africa.

Much like other European explorers, Magellan was not a scientist nor was his expedition a scientific endeavour. Nevertheless, it was science and not just boldness and luck that underpinned the first passage from the Atlantic to the Pacific, the first Pacific crossing, and the first circumnavigation, all in one sitting. We often associate major breakthroughs to a handful of geniuses, but there were none in this story. Instead, it was the slow but steady work of many which made Magellan's voyage possible. Some of these contributors were well-learned, while others were artisans who could barely read. Many remain anonymous or mere footnotes in history: the sailors who provided sailing directions, the cartographers that turned them into charts, the craftsman who built compasses and quadrants, the pilots that used them, the mathematicians who computed almanacs, and the astronomers that consulted them. The (R)evolution of Magellan came from all of them.

Lurking beneath the excitement of the *Victoria*'s return, reservations simmered. Three ships and the lives of two-thirds of the crew, including the captain-general, had been lost. Judging by the worrisome news of conflict

---

[12] Idem, on p. CCCXIX.

and violence brought earlier by the deserting *San Antonio*,[13] more than foul weather had mowed down the crew numbers. The rights of ownership of the Moluccas remained unassigned. The expedition's most obvious and immediate consequence, proving the Earth was circumnavigable, had not been planned and broke the Treaty of Tordesillas.[14] And what would turn to be its furthest-reaching implication, the colonisation of the Philippines, was not yet a clear opportunity.

Aware of the importance of steering the narrative, Elcano wrote a letter to the king immediately after the *Victoria* arrived to Sanlúcar de Barrameda. The memo provided a brief account of the three-year expedition, focusing on gains rather than plights: the discovery of the strait, the crossing of the vast Pacific Ocean, mankind's first circumnavigation, peace agreements signed with the natives, a hull full of cloves, plus samples of further riches (gold from the Philippines, mace and nutmeg from Banda, sandalwood from Timor, pepper, cinnamon, camphor, ginger, and pearls from a myriad of other places).[15]

As the Holy Roman Emperor, Charles V had much on his mind but nevertheless responded promptly: Elcano and a few men of his choosing were to meet him at once. Pigafetta was not among Elcano's entourage, but he also travelled to Valladolid seeking a word with the king. A packed court awaited, eager for details about the voyage. Over the course of several weeks, the crew was interviewed at length by the king, his advisors, and other courtiers. Few were left indifferent by what these men had endured or what they had accomplished. Magellan's contributions were less clear, though. Elcano painted the fallen captain-general as a despotic leader with little regard for the king's orders or his crew's safety.[16] Pigafetta, on the other hand, rarely faulted the fleet's "so noble a captain".[17]

It seems Charles V and his advisors sensed the truth to be somewhere in the middle, and that it would be folly to try to place it accurately. By dragging the matter further, they also risked casting a shadow of dissent and disorder over the entire expedition, prompting Spain to lose face with its European

---

[13] See A Bad Ending to a Great Start.

[14] See How Did Magellan Convince the Spanish King?

[15] "Carta de Juan Sebastián Elcano sobre su viaje de circunnavegación o primera vuelta al mundo" (Seville: Archivo General de Indias, ES.41091.AGI//PATRONATO,48,R.20, 1522), https://pares.mcu.es/ParesBusquedas20/catalogo/description/7343174.

[16] "Información recibida por el alcalde de casa y corte, Santiago Díaz de Leguizamo, en que declaran el capitán de la nao Victoria, Juan Sebastián Elcano, con Francisco Albo y Fernando de Bustamente, sobre distintos pormenores del viaje de la primera vuelta al mundo" (Seville: Archivo General de Indias, ES.41091.AGI//PATRONATO,34,R.19, 1522).

[17] Antonio Pigafetta in James Alexander Robertson, *Magellan's voyage around the world* (Cleveland: The Arthur H. Clark Company, 1906), on p. 177.

peers and providing Portugal with leverage for the upcoming diplomatic clash over the ownership of the Moluccas. Elcano and the other Spanish officers were exonerated from charges of mutiny, and the earlier decision not to prosecute the crew of the *San Antonio* for desertion was reaffirmed. Mesquita, Magellan's cousin brought in chackles aboard the *San Antonio*, was freed from jail and the charges of acts of cruelty at San Julián[18] dismissed.

For some, rewards were in order. The *Victoria's* officers received perpetual pensions, with Elcano's amounting to the handsome sum of 500 ducats a year (187,500 maravedies, more than Magellan's salary of 146,000 maravedies). With the notable exception of most of the *San Antonio* deserters, both dead and alive crew members received the salary accrued during the 3-year expedition. *Quintaladas*, a share of the proceeds from the sale of cloves which easily surpassed the base salary,[19] was paid to about 40 men (the *Victoria's* crew members on its final journey from the Moluccas, plus a few others who had played an important role in securing the cargo, and minus those that had been committed crimes or hidden their true nationality to circumvent the expedition's quota on Portuguese nationals).[20]

Still, one thing was to have an amount due written on a piece of paper, and another was to receive the money. Pigafetta seems to have been one of the first to successfully collect part of his fees, while at Valladolid for his court hearing. Others, particularly the families of the dead sailors, were not so fortunate and a slow trickle of payments lasted at least till 1535.[21] Some salaries were never paid, including Magellan's. His wife and son had died before the *Victoria* returned to Seville, and his siblings never came forward to claim his brother's rights.[22] For most, such returns hardly seemed worthy of a 2-in-3 chance of dying.

Among the surviving one-third (Fig. 17.3), the 55 deserters from the *San Antonio* and the 18 men that returned with Elcano aboard the *Victoria* were luckier than the others. The 12 sailors[23] stranded at Cape Verde were thrown in a Portuguese jail and needed a few more months to finally return home. Worse off still were the survivors from the *Trinidad* (a mere 4 out of the

---

[18] See The Uprising.

[19] See What Drove Men to a Life at Sea?

[20] "Informaciones sobre sueldos, mercancías y mercedes relativas a la Armada a la Especiería organizada por Fernando de Magallanes" (Seville: Archivo General de Indias, ES.41091.AGI/17/ /CONTADURIA,425,N.1,R.1, c.1524), https://pares.mcu.es/ParesBusquedas20/catalogo/description/ 7344884.

[21] Idem.

[22] Tim Joyner, *Magellan* (Maine: International Marine, 1992), on p. 240.

[23] Plus Manuel, a Moluccan.

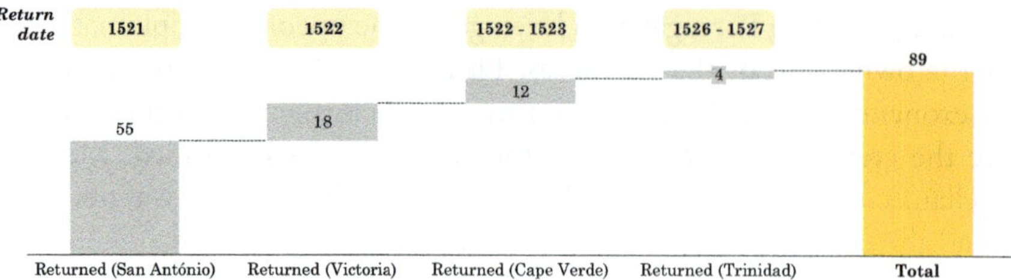

Fig. 17.3 Tally of survivors[24]

25 captured by the Portuguese in the Moluccas[25]), who suffered for years before returning. Thousands of kilometres away from home, far from the strained yet tactful relations of King Manuel I and King Charles V, these men were left to the whims of local governors. Espinosa (the master-at-arms who helped Magellan curb the San Julián mutiny), Juan Rodríguez (deaf and, at 48, one of the oldest sailors on the fleet), and Ginés de Mafra (who later would author one of the first-hand accounts of the expedition[26]) were taken to Malacca and locked up. León Pancaldo (the author of one of the voyage's logbooks[27]) remained at the Moluccas for interrogation before joining his comrades. From there, these four survivors, plus a few others who had not succumbed to the sickness-infested Malaccan cells, were shipped to Ceylon and then to Cochin. Here, the haggard and hungry prisoners were sent to the streets, and soon found themselves begging for subsistence. Vasco da Gama, by then Portuguese India's viceroy, forbade Portuguese captains to take the mariners back to Europe. Somehow, in 1525, Rodríguez was able to board a ship to Lisbon, where he was promptly arrested and only released in 1526. Pancaldo, helped by Genoese kinsmen, hid inside another ship, but was discovered mid-way and marooned in Mozambique. The resourceful sailor sneaked aboard a second convoy and sailed to Lisbon. As Rodríguez, he was jailed but eventually freed, and returned to Spain in 1527. Espinosa and Mafra, who had remained at Cochin, were transferred to a Lisbon prison in 1526, and returned home in 1527.[28]

---

[24] Adapted from: Tim Joyner, *Magellan* [...], on p. 264, 265; Guadalupe Fernández Morente, "Los hombres de la primera vuelta ao mundo", *Carlos y el mar: el viaje de circunnavegación de Magallanes-Elcano y la era de las especias*, 2021, on p. 142.

[25] Twenty-four survivors from the Trinidad, plus the 5 men left at the Moluccas to oversee the trading post (see The Eastern Dead-End).

[26] See How Do We Know What Happened in the Expedition?

[27] Idem.

[28] Tim Joyner, *Magellan* [...], on p. 241–243.

Further hardships also awaited the *Victoria*. While some believe it was spared from extra duty and kept on a dock for all to marvel at the vessel which had first circumnavigated the globe, documents suggest otherwise. Apparently it was repaired, put back on the Indies route, and eventually sunk on a return trip from the Dominican Republic.[29] The Monastery of *La Victoria*, where Magellan pledged his loyalty to Spain and where he wished to be buried, was demolished in the nineteenth century. The wooden figure of *Virgen de la Victoria*, in front of which a barefoot Elcano prayed after the return, still exists although no one is quite sure if it is still the same sculpture.[30]

For Spain's treasury, the mission was hardly a success either. If more ships had returned to Seville, the Crown and the expedition's private investors would have turned a tidy profit. Without it, the proceeds of the cloves brought back by the *Victoria*, some 9 million maravedies,[31] were not enough to cover the fleet's total cost of 10 million maravedies.[32]

In the following years, retracing Magellan's route would prove hazardous, both politically and navigationally. Shortly after returning, Elcano started to prepare a follow-up mission, but complaints from the Portuguese king delayed the enterprise. A fleet of seven ships finally set sail in 1525, led by Loaísa and with Elcano second-in-command. Both officers perished, along with most of their men. Only 24 out of 450 reached the Moluccas, where they were promptly imprisoned by the Portuguese.[33] Further attempts, launched both from Spain and from New Spain's West coast, also failed.

But the tides were slowly changing. In 1529, with neither Portugal nor Spain able to prove the Moluccas fell on their half of the world, an alternative agreement was reached. In the Treaty of Zaragoza Spain sold its claims to Portugal for 131 million maravedies, enough to finance Magellan's expedition thirteen times over.[34] The cost of the follow-up missions was not a write off either, as the additional intelligence gathered aided the later takeover of the Philippines. This would prove to be Spain's most resilient colony, only

---

[29] José Luis Comellas, *La primera vuelta al mundo* (Madrid: Ediciones Rialp, 2012), on p. 145.

[30] Manuel Álvarez Casado in *El Viaje Más Largo* (Madrid: Sociedad Estatal de Acción Cultural, 2019), on p. 291.

[31] Luís Thomaz, *O drama de Magalhães e a volta ao mundo sem querer* (Lisboa: Gradiva, 2018), on p. 68.

[32] Lourdes Díaz-Trechuelo, "La organizacíon del viaje Magallânico", *Actas do II Colóquio Luso-Espanhol de História Ultramarina*, 1975, p. 265–314, on p. 287.

[33] José Maria Moreno Madrid and Henrique Leitão, *Atravessando a porta do Pacífico* (Lisbon: By the Book, 2020), on p. 28.

[34] See What Happened to the Moluccas?

ceded to the United States at the end of the nineteenth century, well after the independence of most of Latin America.[35]

Meanwhile, other European powers were looking intently at what Portugal and Spain were doing. Any hopes the Iberians still harboured about discreetly exploiting overseas opportunities quickly ended after the return of the *Victoria*. Pigafetta, after gifting a handwritten copy of his diary to Charles V, did the same to Louise of Savoy, regent of France. A few years later, he turned his journal into a book,[36] which circulated through Europe in several editions. Maximilianus Transylvanus' account of the expedition,[37] based on the testimonies of the *Victoria* survivors and published in 1523, also spread like wildfire. Beyond the confines of the old continent, a whole new world of unexplored resources awaited. From this point onward, only existing alliances stood in the way of an all-out land grab.

France was the first to react. It had recently fought the 1508–1516 War of the League of Cambrai against Spain and the Papal States, so held little respect for the Treaty of Tordesillas. In 1524, France commissioned a Florentine corsair named Giovanni da Verrazzano to find a Northern American passage to the Indies.[38] He failed, and so did subsequent explorers at the service of France, but the attempts were the first steps towards the country's later control over large swaths of North America, particularly in Canada and Louisiana (the latter named in honour of Louis XIV).

Britain, apart from financing John Cabot's 1497 expedition to North America, initially respected Spain's expansion, as the two countries had fought alongside on the War of the League of Cambrai. However, as Charles V grew increasingly powerful both in Europe and outside it, alliances shifted, and Henry VIII supported France in the 1526–1530 War of the League of Cognac. The British king grew further apart from the Holy Roman Emperor by initiating the separation of the Church of England from the Roman Catholic Church. His daughter, Elizabeth I, concluded the separation and severed the lingering ties to Spain. The Queen's first venture into maritime expansion was the financing of corsair attacks to Spanish and Portuguese ships, to gain a foothold in the Atlantic slave trade.[39] Attacks to Spain's New World possessions followed, with privateers such as Francis Drake and

---

[35] See What Happened to the Philippines?

[36] Antonio Pigafetta in H.E.J. Stanley, *The first voyage round the world, by Magellan. Translated from the accounts of Pigafetta, and other contemporary writers* (London: Hakluyt Society, 1874), on p. 35–163.

[37] Transylvanus in H.E.J. Stanley, *The first voyage round the world* [...], on p. 179–210.

[38] Lawrence Wroth, *The Voyages of Giovanni da Verrazzano* (New Haven: Yale University Press, 1970), on p. 15, 16.

[39] Basil Morgan, "Hawkins, Sir John (1532–1595), merchant and naval commander", *Oxford Dictionary of National Biography*, retrieved January 2025, doi.org/10.1093/ref:odnb/12672.

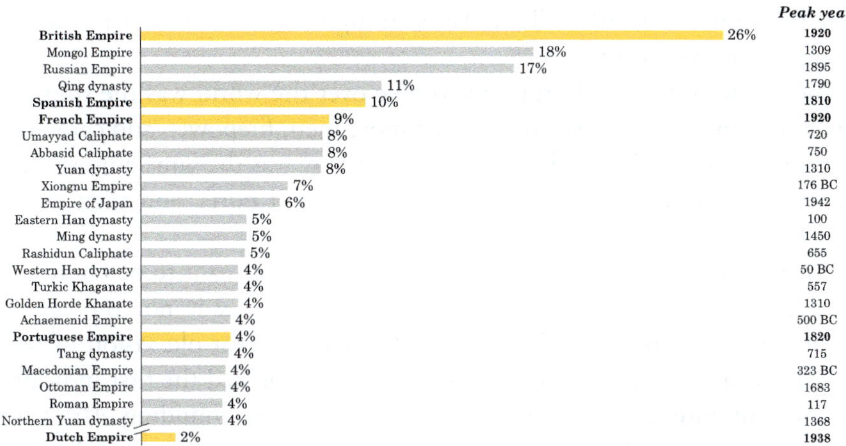

Fig. 17.4 Top 25 largest empires (plus the Dutch Empire), by percentage of world surface area[40]

Thomas Cavendish using the Strait of Magellan to attack Pacific coast settlements. Britain was not the first global European empire but it would grow to be the largest one, eventually spreading to a quarter of the Earth's land area and population (Fig. 17.4).

The Netherlands joined the game late but quickly made up for lost time. In 1566, a Protestant faction rebelled against Spanish rule and took the chance to expand overseas. The Dutch focused on the Southeast Asia islands, first expelling the Portuguese from Malacca and the Moluccas, and then expanding to present-day Indonesia.

The European expansion was a storm of global proportions. The New World, initially just a nuisance in the way of getting to Asia, quickly became the eye of the cyclone. The South American tribes encountered by Magellan's fleet no longer exist or have been reduced to a rounding error. Further up, in Northern and Central America, those encountered by Columbus and other explorers met a similar fate. After European contact, the Native American population decreased by some 70%, from disease (smallpox in particular), war, hardship, acculturation, and miscegenation (both with Europeans and

---

[40] Adapted from: Rein Taagepera, "Expansion and Contraction Patterns of Large Polities: Context for Russia", *International Studies Quarterly*, 41(3), 1997, p. 492–504, on p. 482–484, doi:10.1111/0020-8833.00053; Rein Taagepera, "Size and Duration of Empires: Growth-Decline Curves, 600 BC to 600 AD", Social Science History. 3 (3/4), 1979, p. 115–138, on p. 118, doi:10.2307/1170959; Peter Turchin, "East–West Orientation of Historical Empires", *Journal of World-Systems Research*, 12(2), 2006, p. 222–228, on p. 222, doi:10.5195/jwsr.2006.369; Sebastian Conrad, "The Dialectics of Remembrance: Memories of Empire in Cold War Japan", *Comparative Studies in Society and History*, 56(1), 2014, p. 4–33, on p. 8, doi:10.1017/S0010417513000601

with the more than 12 million slaves brought from Africa to replace the dwindling native workforce).[41]

The key tool used by Europeans to take over the world was the ship. The nau, and later the larger galleon, carried more men, firepower, and cargo than anything else that roamed the seas.[42] The strategy was the same almost everywhere: first take over the major trading routes and ports, and then expand inland.

The hegemony of the ship first required all major oceans to be connected. For centuries, Europeans had taken for granted this was not the case: in the absence of dispelling evidence, Ptolemy's 1,300-year-old vision of land-locked oceans stood. In 1488, Bartolomeu Dias provided the first indication the Ancient Greeks might have gotten it wrong by rounding Africa and demonstrating how the Atlantic and Indian Oceans were connected. In 1492, Columbous stumbled into a new continent unknown to Ptolemy. In 1522, the return of the *Victoria* showed the Atlantic Ocean was connected to a new Pacific Ocean which in turn blended into the Indian Ocean, meaning the entire world was navigable.

These findings quickly found its way onto nautical charts. One of the earliest examples is a c.1521 polar chart[43] made by Pedro Reinel. It contains the first known representations of the coastline south of the La Plata River and of the Falkland Islands, plus the entrance—but not the exit—of the strait. This suggests the chart was made only with the intelligence brought back by the *San Antonio,* which deserted before the fleet fully surveyed the strait. The entire passage to the Pacific makes its debut on the 1523 Turin planisphere, based on the reports from the *Victoria*'s crew. These reports also informed Nuño García de Toreno's 1522 planisphere, which includes the few islands of the Philippines visited by the fleet. Toreno, like all ensuing Spanish cartographers, positioned the Moluccas inside the Spanish hemisphere. Portuguese map makers, on the other hand, placed the islands on their own hemisphere

---

[41] See What Happened to the Natives Encountered by the Fleet?

[42] See How Advanced Were Magellan's Ships?

[43] Unlike the period's more common rectangular charts centred in Europe and Africa, a polar chart is made of two circles, one centred on the North Pole and the other on the South Pole.

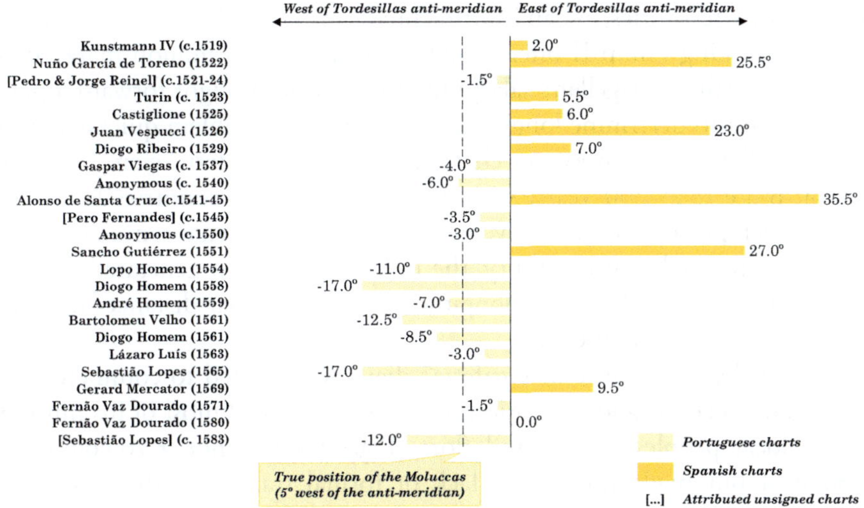

Fig. 17.5 Position of the Moluccas relative to the Tordesillas anti-meridian in Portuguese and Spanish 16th-century charts[46]

(Fig. 17.5). This quarrel dragged on for decades, even after the 1529 Treaty of Zaragoza, perhaps because this agreement included a nullification clause that would force Spain to give back the money if the islands were found to be west of the anti-meridian.[44] Squabbles aside, it is questionable whether Portugal or Spain ever truly knew the precise longitude of the Moluccas.[45]

This issue also plagued navigation. Being able to measure latitude but not longitude was sailing with one eye open and one eye closed. It had not been too serious of a problem for Bartolomeu Dias and his kinsmen when going up and down the African coastline, but it now stood in the way of global navigation, made predominantly in the east–west direction. Not only was measuring longitude at sea all but impossible, there was not even a way to dependably

---

44 Emma Blair and James Robertson, *The Philippine Islands* (Cleveland: A.H. Clark Company, 1903), in Volume 1, p. 227, 228.

45 In 1610, João Baptista Lavanha, a Portuguese mathematician, still complained that astronomers capable of accurately determining the longitude of the Moluccas were not interested in stepping onto a ship and travelling to the Indies (João Baptista Lavanha in Francisco Mendes da Luz, *O Conselho da India: contributo ao estudo da administração e do comércio do Ultramar Português nos princípios do século XVII* (Lisboa: Agência Geral do Ultramar, 1952), on p. 165, 166).

46 Adapted from Joaquim Gaspar and Šima Krtalić, *The Cartography of Magellan* (Lisbon: Tradisom, 2023), on p. 242.

position new territories, as San Martín's predicaments had showed.[47] In 1567, the Spanish king Philip II offered 6,000 ducats[48] (2.25 million maravedies, 15 years' worth of Magellan's salary) to anyone who brought forward a practical solution to determine longitude. The longitude problem quickly turned into one of the most important challenges for early modern science. Solving it took more than 200 years, with multiple European countries launching prizes which prompted the involvement of some of the Renaissance's brightest minds. By the eighteenth century, the fierce competition generated not one but two methods to determine longitude at sea: the lunar distances method, and the marine chronometer.

Nevertheless, very little of Pigafetta's diary is dedicated to commerce and politics, or to charts and navigation. Instead, he devoted most of his musings to the local peoples, fauna, and flora. Pigafetta was not a zoologist nor a botanist but made up for the lack of formal training with unbounded creativity. In the Atlantic crossing, he described a *"kind of bird which only lives on the droppings of the other birds"*[49] (parasitic jaegers, which force others to regurgitate freshly captured fish). In Brazil, he marvelled at furry yellow "*cat-monkeys*"[50] (golden lion tamarins, monkeys which indeed resemble felines) and *"pigs which have their navel on the back"*[51] (peccaries, pig-like creatures with a scent gland that looks like a navel). In Patagonia, Pigafetta observed large colonies of black fat "geese"[52] incapable of flying (the first description of Magellanic penguins). During those long winter months he documented many other species, including South American sea lions, greater rheas, culpeos, and guanacos. The description of the latter is unceremonious but remarkably accurate: *"This beast has its head and ears of the size of a mule, and the neck and body of the fashion of a camel, the legs of a deer, and the tail like that of a horse, and it neighs like a horse"*.[53] Sometimes though, the fledging biologist got a bit carried away. While in Southeast Asia, he penned the first known description of leaf insects: *"In this island are also found certain trees, the leaves of which, when they fall, are animated, and walk. [...] If they are touched*

---

[47] See Were San Martín's Longitude Measurements Accurate?

[48] Richard de Grijs, "European Longitude Prizes. I. Longitude Determination in the Spanish Empire", *Journal of Astronomical History and Heritage*, 23, 2020, p. 1–32, on p. 9, doi.org/10.48550/arXiv.2009.12778.

[49] Antonio Pigafetta in H.E.J. Stanley, *The first voyage round the world, by Magellan* [...], on p. 42.

[50] Idem, on p. 46.

[51] Idem.

[52] Idem, on p. 49.

[53] Idem, on p. 50.

they escape, but if crushed they do not give out blood. I kept one for nine days in a box. When I opened it the leaf went round the box. I believe they live upon air".[54]

Pigafetta's descriptions of local peoples and their customs are equally vivid. But, more than the occasional tall tale of colossal Patagonians or homunculus with giant ears, what stands out most from these portrayals is their neutral stance. Rarely does the rapporteur pass judgement and, when he does, it is usually to entice a smirk from the reader. At San Julián, amidst the cold and the storm of mutiny, he finds the time to describe the unusual distribution of labour among the Patagonians: "*Then these men came, who carried only their bows in their hands; but their wives came after them laden like donkeys, and carried their goods*".[55] At Guam, Pigafetta describes the Chamorros not as heathens but matter of factually as "*people liv[ing] in liberty and according to their will, for they have no lord or superior*". After Magellan ordered a raid to kill natives and burn their huts as punishment for stealing,[56] the Italian lamented their suffering ("*when we wounded any of this kind of people with our arrows, which entered inside their bodies, they looked at the arrow, and then drew it forth with much astonishment, and immediately afterwards they died, which did not fail to cause compassion*") and understands their grief ("*we saw some of these women, who cried out and tore their hair, and I believe that it was for the love of those whom we had killed*"[57]).

Transylvanus, too, reinforced the humanistic significance of the expedition:

For who can believe that these were Monosceli, Scyopodæ, Syrites, Spitamei, Pygmies, and many others, rather monsters than men. And as so many places beyond the Tropic of Capricorn have been sought, found, and carefully examined, both by the Spaniards in the south-west and by the Portuguese sailing eastwards, and as the remainder of the whole world has now been sailed over by our countrymen, and yet nothing trustworthy has been heard concerning these man-monsters, it must be believed that the accounts of them are fabulous, lying, and old women's tales, handed down to us in some way by no credible author[58]

---

[54] Idem, on p. 119.

[55] Idem, on p. 51.

[56] See Bad News.

[57] Antonio Pigafetta in H.E.J. Stanley, *The first voyage round the world, by Magellan* [...], on p. 69.

[58] Transylvanus in H.E.J. Stanley, *The first voyage round the world, by Magellan* [...], on p. 185.

So then, Magellan's expedition mattered enormously. On the one hand, it contributed to advance navigation, astronomy, cartography, anthropology, and biology. On the other hand, it was a key log in European colonialism, which shaped many of the world's current borders, demographics, cultures, and religions.

Did it matter for Magellan, though? He died taking one risk too many,[59] and was soon followed by his wife and son. He had severed ties with the remaining family in Portugal, leaving no one to carry the family's coat of arms. He never became rich, nor did he reach the status he longed for. The Portuguese king called him a traitor, and the Spanish monarch seemed ambivalent about the explorer. Yet his name lived on.

In the Philippines, Magellan's ambiguous standing is not unlike that of Columbus in the Americas. There are three cities called *Magallanes*, plus a score of street names and other markers dedicated to the expedition. At Cebu, Magellan's Cross Pavilion houses what is thought to be the same Christian Cross erected by the crew, which is a popular pilgrimage destination for the 86 million Catholic Filipinos (close to 80% of the population).[60] At Mactan, the place where the captain-general fell is called Magellan Bay and hosts the Magellan Monument. Yet, just next to it, there is a statue of Lapulapu, who led the defence force that killed Magellan. In 2021, as others were preparing to celebrate the 500 years of the first circumnavigation, Filipinos honoured the natives who fed the starving European crew, and Lapulapu's resistance to occupation.[61]

Elsewhere in the world, the explorer's name lives on at the Strait of Magellan,[62] the Magellanic penguin (*Spheniscus magellanicus*, the Patagonian "geese" described by Pigafetta), and the Magellanic clouds (two dwarf galaxies described by Pigafetta[63] but known since ancient times). Other uses mark discoveries which go well beyond anything that Magellan and his peers had anything to do with or even imagined possible: *Magelhaens*, a Lunar crater

---

[59] See Worse News.

[60] "Religious Affiliation in the Philippines (2020 Census of Population and Housing)", Philippine Statistics Authority, 2020, https://psa.gov.ph/content/religious-affiliation-philippines-2020-census-population-and-housing.

[61] Kate Fullagar and Kristie Flannery, "Ferdinand Magellan's death 500 years ago is being remembered as an act of Indigenous resistance", *The Conversation*, 2021, https://theconversation.com/ferdinand-magellans-death-500-years-ago-is-being-remembered-as-an-act-of-indigenous-resistance-158226.

[62] Originally called *Canal de Todos los Santos* (Channel of All Saints), but nautical charts as early as 1525 already referred to the Strait of Magellan e.g., Diogo Ribeiro's Castiglione planisphere (Joaquim Gaspar and Šima Krtalić, *The Cartography of Magellan* [...], on p. 168–180).

[63] "*The Antarctic Pole is not so starry as the Arctic. Many small stars clustered together are seen, which have the appearance of two clouds with little distance between them, and they are somewhat dim*" (Antonio Pigafetta, *The first voyage around the world 1519–1522*, edited by Theodore Cachey Jr., Toronto, University of Toronto Press, 2007, on p. 25).

and a Martian one; 4055 Magellan, an asteroid; the Giant Magellan Telescope; and NASA's Magellan Venus probe. Why is Magellan's name pegged to these scientific achievements? He was not a scientist, nor was his expedition a scientific endeavour.

Yet it was science, and not only boldness and luck that underpinned the first passage from the Atlantic to the Pacific, the first Pacific crossing, and the first circumnavigation, all in one sitting. There are no prominent geniuses in this story (Kepler, Galileo, and Newton were not even born), nor is an *annus mirabilis* anywhere to be found.[64] Instead, it was the work of many that underpinned the scientific achievements behind Magellan's voyage—and behind what happened next. Some of these contributors were well-learned, while others were artisans that could barely read. Many remain anonymous or a mere footnote in history: the sailors who provided sailing directions, the cartographers that turned them into charts, the craftsman who built compasses and quadrants, the pilots that used them, the mathematicians who computed almanacs, and the astronomers that consulted them.

This process was a slow but steady one, made of small steps rather than giant leaps. One can argue that calling this book *The Revolution of Magellan* instead of *The Evolution of Magellan* is succumbing to Pigafetta's literary panache. Then again, not all revolutions happen quickly.

---

[64] In 1905, Albert Einstein published four breakthrough papers, describing the photoelectric effect, the Brownian motion, the theory of special relativity, and the principle of mass-energy equivalence.

9783032107961